Single and Cross-Layer MIMO Techniques for IMT-Advanced

RIVER PUBLISHERS SERIES IN COMMUNICATIONS

Volume 13

This series focuses on communications science and technology. This includes the theory and use of systems involving all terminals, computers, and information processors; wired and wireless networks; and network layouts, procontentsols, architectures, and implementations.

Furthermore, developments toward new market demands in systems, products, and technologies such as personal communications services, multimedia systems, enterprise networks, and optical communications systems.

- Wireless Communications
- Networks
- Security
- Antennas & Propagation
- Microwaves
- Software Defined Radio

For a list of other books in this series, see final page.

Single and Cross-Layer MIMO Techniques for IMT-Advanced

Filippo Meucci

University of Florence, Italy

River Publishers
Aalborg

ISBN 978-87-92329-50-9 (hardback)

Published, sold and distributed by:
River Publishers
P.O. Box 1657
Algade 42
9000 Aalborg
Denmark

Tel.: +45369953197
www.riverpublishers.com

Contents

II Cross-Layer MIMO

Preface

In the last two decades, the wireless arena has witnessed an astonishing number of technologies playing a role in the definition of new wireless systems. Driven by the pressing capacity demand, the research community has developed several technological enablers. Some of them have been fundamental for laying the groundwork to a performance leap; the impact of the others have been sometimes overestimated.

In this complex panorama, technological blocks that will be part of wireless systems in the near-future surely include: Orthogonal Frequency Division Multiplexing (OFDM) modulation at the physical (PHY) layer, Multiple Input Multiple Output (MIMO) systems, and a CL stack design. While the benefits of OFDM have been recognized for several years, the real capacity improvement of MIMO antenna is still being debated today. As to the last point, even if opportunities for CL have been pointed out for a long time, the impact on the actual legacy systems has not been noticeable, cooled down from the threats that network managers and investors see in a system design paradigm shift.

This monograph will present some advanced MIMO techniques to be used with adaptive mechanisms over a single layer and a MIMO-ARQ protocol designed for best exploiting MIMO in conjunction with CL operation. Adaptivity, CL approach, and MIMO antenna are analyzed together to show a deep impact on the sum-capacity achievable over the wireless link.

With no loss of generality, the designs proposed in this monograph are evaluated inside an IEEE802.16 wireless network with Orthogonal Frequency Division Multiple Access (OFDMA) at the PHY layer and a connection-oriented MAC. The technical solutions are directly applicable for Long Term Evolution (LTE) systems, which share the same conceptual PHY and MAC configurations: the choice of OFDMA and connection-oriented MAC layer allows to evaluate the impact of MIMO techniques inside the roadmap for future International Mobile Telecommunications Advanced (IMT-A) systems. In particular, the cross-layer MIMO-ARQ design will provide an additional support in order to achieve target peak data rates of up to approximately 100

Mbit/s for high mobility terminals and up to approximately 1 Gbit/s for low mobility terminals, as requested in the IMT-A specifications published by International Telecommunication Union (ITU).

The introduction presents the functional requirements for IMT-A candidate systems and the relation between IEEE802.16 and LTE wireless access networks. In the first part of the monograph, adaptive strategies are analyzed separately at the PHY and MAC layers. Two solutions are proposed at the PHY layer. The first one provides variable spectral efficiency via adaptive MIMO scheme selection based on upper-bounded Bit Error Rate (BER). MIMO transmission is varied among full Spatial Multiplexing (SM) and full diversity schemes, also including a trade-off between the two. A second adaptive strategy selects between Space Time Block Code (STBC) and Space-Frequency Block Coding (SFBC) transmission depending on the channel delay spread and Doppler shift. At the MAC layer, Automatic Repeat Request (ARQ) and Dynamic Service Addition (DSA) protocols are analyzed. The length of ARQ retransmissions is optimized jointly with adaptive Modulation and Coding Schemes (MCS). Then, the probability of signaling and admission blocking of the DSA protocol is expressed in closed form based on the parameters of the three-way handshake protocol.

The second part of the monograph presents an evolution of the previous approach, providing a CL MIMO-ARQ protocol, where adaptive MIMO schemes, namely SM and Alamouti STBC, are used with ARQ protocol. A Multiple User (MU) network is served in DownLink (DL) with a Round Robin (RR) scheduler; the design is ready to include more advanced schedulers. The ARQ state machine at the MAC layer is aware of per-antenna ACK. The interaction between the ARQ and the PHY layer, with a per-antenna ACK layer, allows resource exploitation to increase shifting

from *MIMO Signal Processing Gain* to *MIMO Protocol Gain*

with no need for Channel State Information (CSI) feedback. The absence of CSI feedback at the PHY layer is an important characteristic of the proposed MIMO-ARQ CL designs since MIMO CSI feedback (when feasible) drastically reduces the network efficiency.

The added degrees of freedom offered by MIMO-ARQ CL transmissions make a difference in terms of capacity if correctly exploited in particular for overcoming the problem of low MIMO channel ranks. The advantages of the paradigm shift from signal processing gain to protocol gain – together with the modifications to be applied at the classical protocol stack – are discussed in the final chapter.

The author would like to thank many colleagues who have shared their visions and ideas and helped to construct a deep insight into the problems and techniques presented in this work. First of all, Professors Ramjee Prasad and Enrico Del Re, who gave me the opportunity to be part of very lively research centers, respectively CTIF, Center for TeleInFrastruktur in Aalborg, Denmark, and the LENST Laboratory at the University of Florence, Italy.

I would also like to thank Dr. Laura Pierucci, Dr. Albena Mihovska, Dr. Elisabeth De Carvalho, and Dr. Petar Popovsky for their invaluable technical discussions.

Filippo Meucci
September 2010

List of Abbreviations

1G	First Generation
1xEV-DO	1× Evolution-Data Only
2G	Second Generation
3G	Third Generation
4G	Fourth Generation
AAS	Adaptive Antenna Systems
AC	Admission Control
ACK	Acknowledgement
ACL	Adaptive Cross Layer
AMC	Adaptive Modulation and Coding
AN	Analytically
ARQ	Automatic Repeat Request
BE	Best Effort
BER	Bit Error Rate
BPSK	Binary Phase Shifting Key
BS	Base Station
BSN	Block Sequence Number
CDF	Cumulative Distribution Function
CDI	Channel Distribution Information

CDMA	Code Division Multiple Access
CID	Connection IDentifier
CL	Cross Layer
COID	Co-Ordinated Interleaved Orthogonal Designs
CP	Cyclic Prefix
CQI	Channel Quality Indicator
CRC	Cyclic Redundancy Check
CSI	Channel State Information
CSIR	CSI at the Receiver
CSIT	CSI at the Transmitter
DCD	Downlink Channel Descriptor
DFT	Discrete Fourier Transform
DHCP	Dynamic Host Control Protocol
DL	DownLink
DLC	Data Link Control
DLFP	DownLink Frame Prefix
DPC	Dirty Paper Coding
DSA	Dynamic Service Addition
ED	Euclidean Distance
ertPS	extended rtPS
FD	Frequency Domain
FDD	Frequency Division Duplex
FDMA	Frequency Division Multiple Access

FSMC	Finite State Markov Chain
FTP	File Transfer Protocol
FUSC	Fully Used SubCarriers
GMH	General Mac Header
HARQ	Hybrid Automatic Repeat Request
HOL	Head Of Line delay
HSXPA	High-Speed Downlink/Uplink Packet Access
ICI	Inter-Carrier Interference
IDFT	Inverse Discrete Fourier Transform
IEEE802.16	Institute of Electrical & Electronics Engineers 802.16
IFDMA	Interleaved Frequency Division Multiple Access
iid	identically independent distributed
IMT-A	International Mobile Telecommunications Advanced
IP	Internet Protocol
IS-2000	Information System 2000
ISI	Inter-Symbol Interference
ITU	International Telecommunication Union
LLC	Link Layer Control
LoS	Line of Sight
LSTBC	Layered STBC
LTE	Long Term Evolution
LTE-A	Long Term Evolution Advanced
LTV	Linear Time Variant

MAC	Medium Access Control
MC-CDMA	Multi Carrier Code Division Multiple Access
MCS	Modulation and Coding Schemes
MGF	Moment Generating Function
MIMO	Multiple Input Multiple Output
MISO	Multiple Input Single Output
ML	Maximum-Likelihood
MMSE	Minimum Mean Square Error
MT	Mobile Terminal
MU	Multiple User
NET	Networking
nrtPS	non-real-time Polling Service
NU	New User
OFDM	Orthogonal Frequency Division Multiplexing
OFDMA	Orthogonal Frequency Division Multiple Access
OSIC	Ordered Successive Interference Cancellation
PAM	Pulse Amplitude Modulation
PAPR	Peak to Average Power Ratio
PC	Packet Combining
PDF	Probability Density Function
PDP	Power Delay Profile
PDU	Protocol Data Unit
PER	Packet Error Rate

PHY	physical
PMP	Point to MultiPoint
PO	Packet Overhearing
PSK	Phase Shifting Key
PtP	Point to Point
PUSC	Partially Used Subcarriers
QAM	Quadrature Amplitude Modulation
QoS	Quality of Service
QOSTBC	Quasi-Orthogonal Space-Time Block Code
RAT	Radio Access Technology
RMS	Root Mean Square
RR	Round Robin
RRM	Radio Resource Management
rtPS	real-time Polling Service
SC	Single Carrier
SCa	Single Carrier advanced
SC-FDMA	Single Carrier Frequency Division Multiplexing
SDMA	Spatial Division Multiple Access
SDU	Service Data Unit
SE	Spectral Efficiency
SEP	Symbol Error Probability
SER	Symbol Error Rate
SFBC	Space-Frequency Block Coding

SFR	Single Frequency Reuse
SIC	Successive Interference Cancellation
SIM	Simulations
SINR	Signal to Interference Noise Ratio
SIR	Signal to Interference Ratio
SISO	Single Input Single Output
SM	Spatial Multiplexing
SNMP	Simple Network Management Protocol
SNR	Signal-to-Noise Ratio
SON	Self Organizing Networks
ST	Space-Time
STBC	Space Time Block Code
STC	Space-Time Code
STF	Space-Time-Frequency
STFBC	Space-Time-Frequency Block Code
STTC	Space-Time Trellis Code
STTD	Space-Time Transmit Diversity
SU	Single User
SVD	Single Value Decomposition
TD-CDMA	Time Division CDMA
TDD	Time Division Duplex
TDM	Time Division Multiplcxing
TDMA	Time Division Multiple Access

TD-SCDMA Time Division Synchronous CDMA

UCD Uplink Channel Descriptor

UGS Unsolicited Grant Service

UL UpLink

UMTS Universal Mobile Telecommunications System

VBLAST Vertical BLAST

VoIP Voice over IP

WARC92 World Administrative Radio Conference of 1992

WiMAX Worldwide Interoperability for Microwave Access

WRC07 World Radio Conference of 2007

WRC2000 World Radio Conference of 2000

WSSUS Wide Sense Stationary – Uncorrelated Scattering

ZF Zero Forcing

1

International Mobile Telecommunication – Advanced

Currently, traffic demand in wireless communication networks is growing at a very fast pace: data transfers, voice calls, video streaming, and interactive Web 2.0 browsing are requiring high-capacity links with very tight delay requirements. The scarcity of spectrum resources is forcing wireless communication engineers to push for high spectral efficiencies, which is even more challenging when the Mobile Terminal (MT)[1] has a high degree of mobility, as required in future wireless networks.

The research and the industry communities have set the ambitious goal of providing the so-called "Mobile Internet," i.e. access to the network on handheld devices, data traffic, and multimedia streaming; achieving the same user experience available on a cable-connected computer, but on the move. The achievement of a good performance in such a harsh setting requires several improvements. Modifications are needed in all the layers of the communication stack: if, on the one side, the physical layer spectral efficiency must be increased, then on the other side the network architecture also needs to be simplified with backward compatibility with the legacy systems already deployed and the standardization bodies are currently working on finalizing the specifications of Fourth Generation (4G) wireless networks.

Looking back at the history of wireless cellular systems, each migration among the generations was driven by a specific market force. From the First Generation (1G) to the Second Generation (2G), the main goal was to extend the access to mobile communications to a large mass of users, as the 1G network was providing little room to accommodate an increasing amount

[1] Varied terms are used in literature to indicate the mobile equipment in a cellular network. The IEEE standards use Subscriber Station (SS) or Mobile Station (MS). LTE-A prefers to use User Terminal (UT) or User Equipment (UE). In order to keep the nomenclature as general as possible and to highlight a fully mobile equipment, the acronym Mobile Terminal (MT) is used herewith.

of terminals. One of the main technology upgrades was the passage from analog to digital voice transmission with a subsequent smaller bandwidth for supporting the same amount of calls. From 2G to 3G, the technological upgrades provided a good data connection in terms of both capacity and costs, supporting packet switching networking on a network specifically designed for the voice calls. Packet-based services have been introduced on top of a native circuit switched network.

In the near future, 4G systems will support a secure Internet Protocol (IP)-based connection for all the dishomogeneous services provided to users in high mobility. A sensible performance gain is expected compared with the previous generations. Broadband and high mobility support are, thus, the most pressing requirements to fulfill, and high spectral efficiencies have to be achieved with practical system implementation. The technical performance expected of the 4G systems is defined globally by the ITU, which published a set of requirements collected in a series of recommendations that define the IMT-A. Though 4G is a broader term and could include standards outside IMT-A, 4G is generally associated with IMT-A specifications as published in [49].

The data rate targets have been set to 100 Mbit/s for high and 1 Gbit/s for low mobility, and MIMO has a crucial role that has been intensively proposed for the PHY layers of 4G systems.

1.1 IMT-A Candidates

Although the technical specifications of the upcoming 4G networks are well defined, there are many questions regarding which will be the standard that will lead the commercial deployment of 4G infrastructures and devices. Two big standardization bodies are working to submit a candidate for IMT-A evaluation process: the data-centric Worldwide Interoperability for Microwave Access (WiMAX) and the voice-centric Long Term Evolution Advanced (LTE-A). Although it has a different vision, the latter evolves from its Third Generation (3G) predecessors; by contrast, WiMAX, which was started before LTE-A, adopts a clean-slate approach. The two standards have profound technology similarities and are based on OFDM and heavily exploit MIMO antenna technology to attain very high spectral efficiency at the PHY layer.

On the opposite side, each standard has its peculiar focus, and hence, they can be seen as non-totally competing but partially complementary technologies. The commercial success or predominance of one against the other should not be forecasted on their technical specifications but rather should take into consideration the combination of technology performance, supported services and applications, deployment coverage and requested initial investments.

In this monograph, the performance evaluations have been carried out in the framework of WiMAX, following the specifications of the Institute of Electrical & Electronics Engineers 802.16 (IEEE802.16) standard and referring in particular to IEEE802.16e and IEEE802.16-2009, which includes high mobility configuration profiles. Nonetheless, what is proposed here is applicable also in LTE-A. This is possible since the two PHY layers share the same space-time-frequency-antenna resource allocation mechanisms and both MACs are connection-oriented.

The overall IMT-A requirements, as to the date of publication, are briefly discussed in this section. It is important to note that these requirements are being enhanced in line with the technology developments. ITU compiles the following performance levels based on the ongoing activities and including the latest technical developments.

It is worth mentioning that the system specified in the ITU-R M.1457 recommendation (3G-IMT 2000 Release 1) was already including the 3GPP LTE Release 8 and the IEEE 802.16e-2005 ("Mobile WiMAX"). In that recommendation, many of the main 4G requirements were already specified: support of fixed and mobile terminals supported, all-IP network, flat network architecture, OFDMA, MIMO, Hybrid Automatic Repeat Request (HARQ) at the PHY layer, just to cite a few.

This means that both Mobile WiMAX and LTE are officially designated 3G technologies evolving to 4G inside their respective standards bodies. The race to 4G has been accelerated by a number of factors; the early deployment of Mobile WiMAX is one of the most important.

Since its kick-off in late 2005, WiMAX has been deployed by almost 400 operators spanning 133 countries. Although not an independent source, WiMAX forum predicts a worldwide WiMAX subscriber base of about 130 million by 2012 with a prediction of a strong base of about 27 million subscribers in India. Chip manufacturers like Intel had committed to the availability of embedded WiMAX chips for laptops, PCs and mobile handsets from the end of 2009 on.

The competitor LTE provides an evolution path for advancement from today's 3G Universal Mobile Telecommunications System (UMTS) and 3.5G High-Speed Downlink/Uplink Packet Access (HSXPA) cellular technologies to the future 4G networks. Given its voice-centric vision and its evolutionary path, LTE has been supported by several mobile operators worldwide. This endorsement makes LTE the most likely standard to be dominant worldwide in the short term for mobile voice with seamless roaming across different operators all over the globe. One key advantage of LTE is the possibility to leverage the 700 MHz spectrum for efficient support of voice services with higher in-building penetration and across longer distances.

However, some consider the development road-map of LTE-A too slow (the name itself – Long Term Evolution – is not a guarantee of fastness) and the cellular industry can leverage on LTE as a way of slowing down WiMAX commercial spreading without being committed to providing a real user experience improvement. The advantage of the "evolutionary development" of LTE is also to be verified. In reality, LTE is more a "forklift upgrade" proposing a new Radio Access Technology (RAT), different network architecture and a totally different PHY layer configuration and Base Station (BS) capabilities. LTE modem pricing could be another issue considering an extrapolation of the already available 3G modems: 3G-HSPA modems are priced two to three times higher than a Mobile WiMAX modems delivering a peak rate of 14 Mbit/sec against the 40 Mbit/sec supported in WiMAX.

RAT is one of the key blocks for wireless networks and 4G specifications require high-performing PHY layers with an increases spectral efficiency, capacity and scalability. Time Division Multiple Access (TDMA) and/or Frequency Division Multiple Access (FDMA), used since 1G cellular networks, have their own advantages and drawbacks. In particular, TDMA is less efficient in serving the high data rate connection as it is requiring large guard periods to alleviate the multipath impact while FDMA has low spectral efficiency since more bandwidth is needed for guards to avoid Inter-Carrier Interference (ICI). In the 2G networks, one set of standards used the combination of FDMA and TDMA and another gruop of standards introduced a new access scheme based on Code Division Multiple Access (CDMA). The latter increased the system capacity while being efficient in handling the multipath channel. However CDMA suffers from poor spectrum flexibility, poor spectrum exploitation (due to bandwidth spreading factors) and scalability. CDMA is the base of 3G extensions such as: Information System 2000 (IS-2000), UMTS, HSXPA,1× Evolution-Data Only (1xEV-DO), Time Division CDMA (TD-CDMA) and Time Division Synchronous CDMA (TD-SCDMA).

A big shift in the RATs came with the application of OFDM. Thanks to simple implementation via DFT and IDFT, OFDM has been universally inserted in 4G candidates. OFDM permits very easy channel equalization, efficient signal processing and very flexible MU resource allocation. A variant of pure

OFDM radio access technology is the OFDMA where users can be allocated in parts of the time-frequency transmission frame (resource chunk). Variants of OFDM have been also proposed for reduced Peak to Average Power Ratio (PAPR) such as: Single Carrier Frequency Division Multiplexing (SC-FDMA) and Interleaved Frequency Division Multiple Access (IFDMA). On another side, Multi Carrier Code Division Multiple Access (MC-CDMA) has been developed and chosen for some standards as IEEE802.20 (MBWA).

WiMax uses OFDMA in the DL and in the UL. In LTE-A, OFDMA is used for the DL while SC-FDMA is used in UL. SC-FDMA in LTE-A UL has been considered one of the factors that could lead to a success of LTE over WiMAX due to a reduced PAPR and lower cost BSs. This argument seems not so strong as this issue has been addressed by the research community and solved: several solutions have been presented, the most recent in [110] where, for instance, phase rotations create a constant envelope OFDM signal solving the PAPR problem for non-SC-FDMA uplinks, as in WiMAX.

Based on these considerations, WiMAX has been chosen as the reference standard for the work presented in this monograph. However, the technical solutions proposed here are applicable in both candidates: the RATs share several technical aspects. OFDMA (or its variants as SC-FDMA) are used, user are allocated in portions of the time-frequency grid in the same conceptual way (each user is assigned to a piece of bandwidth for a certain time inside a frame) and identical MIMO schemes are used and, further, both candidates have a connection-oriented MAC layer.

1.2 IMT-A Technical Requirements and Challenges

4G aims at several performance enhancements in respect to the current 3G systems; various areas of research are vibrantly active:

- Higher throughput per sector (both peak rates and average)
- Higher MT speeds supported
- Better coverage with reduced signal fading
- All-IP network
- Flat and simple network architecture
- Native VoIP support
- Lower latency (access < 10 ms and handover < 30 ms)
- Better cell edge throughput
- Enhanced broadcast and multicast services
- Enhanced location based services

- Deployment flexibility – Self Organizing Networks (SON)

Due to spectrum segmentation, operators will have licensed spectrum in widely separated frequency bands with heterogeneous bandwidths. 4G broadband wireless systems should provide a single radio bearer channel with the flexibility to aggregate non-uniform and non-contiguous channels. Dynamic bandwidth allocation and scaling will be an important key for providing the capacity demanded by multimedia services in a dynamic mobile environment. The spectrum will be not anymore a fixed resource but dynamically adapted to traffic and user requirements. MT will be equipped with multiple transceivers operating over several radio channels co-existing with low self-interference (Multi-Radio Multi-Band Terminals). Such technology opens the problem of the added degree of freedom for resources allocation and traffic scheduling. A common Radio Resource Management (RRM) have been proposed in [79].

4G network will be deployed as a Single Frequency Reuse (SFR) network in order to maximize the spectral utilization over large geographical areas. For this reason, interference will degrade the quality of the connections especially for the users at the cell edges with more challenges during UpLink (UL) connections. While in DL a maximum of 8 to 10 interferes are expected, in the UL this number could easily reach more than one hundred. Techniques to manage and reduce interference will be needed jointly with resources and users scheduling. Current 3G networks have two parallel infrastructures for circuit switched and packet switched traffic. Instead, 4G will be based only on packet switching. This requires very low latency times in the control and in the data planes. The IP network in 4G will be supporting native IPv6.

The evolution of smart or intelligent antennas will deeply impact on the future of wireless technologies. Multiple antennas at both sides of the radio link can lead to improved capacity with SM transmission modes. In SM, independent streams can then be transmitted simultaneously from all the antennas increasing the data rate times a number equal to minimum of the number of receive and transmit antennas. The independence of the fading statistic for each antenna is the key for SM. Especially in small devices, low correlation among antenna elements is hard to achieve. Solutions based on antenna selection and Multiple Input Single Output (MISO) systems have been already proposed. Antenna selection provides a sensible cost reduction due to a single transceiver chain attached to a multi-antenna port and the MISO systems allow to have expensive antenna systems concentrated at the BSs, where the antenna elements can be separated several wavelengths achieving very low correlation. MIMO can be exploited not only for increased capacity but also for high reliability for lower error rates and longer-range communications. Reliability is increased with transmit or receive diversity, especially with open-loop Space-Time Code (STC) which does not require the CSI at the Transmitter (CSIT). When CSIT is needed at the transmitter, the transmitted signal can be pre-processed in order to have the best signal at the antenna ports at the MT [34].

1.3 IMT-A Performance Requirements

The performance requirements of IMT-A candidates are analyzed in detail, referring to ITU Recommendation M.2134 [49]. The key features to be fulfilled by the candidate system are:

- a high degree of commonality of functionality worldwide while retaining the flexibility to support a wide range of services and applications in a cost efficient manner
- maintained compatibility inside IMT systems and fixed networks
- capability of interworking with non-IMT radio access systems
- worldwide roaming capability
- high quality mobile services
- user equipment suitable for worldwide use
- user-friendly applications, services and equipment
- enhanced peak data rates to support advanced services and applications.

Quantitative specifications are divided in minimum requirements and target requirements. The data rate for the mobile environment has been set to 100 Mbit/s for high and 1 Gbit/s for low mobility as targets for research. The following are the minimum requirements for the candidates.

1.3.1 Cell Spectral Efficiency

The cell spectral efficiency η is defined as the aggregate throughput of all users (the number of correctly received bits, i.e. the number of bits contained in SDU delivered to layer 3, over a certain period of time) divided by the channel bandwidth divided by the number of cells. The channel bandwidth for this purpose is defined as the effective bandwidth times the frequency reuse factor, where the effective bandwidth is the operating bandwidth normalized appropriately considering the UL/DL ratio.

Table 1.1 IMT-A cell spectral efficiency

Test Environment	Downlink (bit/s/Hz/cell)	Uplink (bit/s/Hz/cell)
Indoor	3	2.25
Microcellular	2.6	1.80
Base coverage urban	2.2	1.4
High speed	1.1	0.7

Let X_i denote the number of correctly received bits by user i (DL or UL) in a system with N users and M cells, ω be the channel bandwidth and T the time over which the data bits are received. The system spectral efficiency, η, is then:

$$\eta = \frac{\sum_{i=1}^{N} X_i}{T \cdot \omega \cdot M} \tag{1.1}$$

For each test environment, as defined in the ITU-R Report M.2135, the following minimum efficiency per cell to be fulfilled by the candidates is reported in Table 1.1.

1.3.2 Peak Cell Efficiency

The peak spectral efficiency is the highest theoretical data rate, which is the received data bits assuming error-free conditions assignable to a single mobile station, normalized by bandwidth when all available radio resources for the corresponding link direction are utilized (i.e., excluding radio resources that are used for PHY layer synchronization, reference signals or pilots, guard bands and guard times). The minimum requirements for peak spectral efficiencies are:

- DL peak spectral efficiency: 15 bit/s/Hz.
- UL peak spectral efficiency: 6.75 bit/s/Hz.

These values were defined assuming an antenna configuration for DL of 4×4 and for UL of 2×4.

1.3.3 Bandwidth

Scalable bandwidth is the ability of the candidate air interface to operate with different bandwidth allocations. This bandwidth may be supported by single or multiple RF carriers. The candidate shall support a scalable bandwidth up to and including 40 MHz even if proponents are encouraged to consider extensions to support operation in wider bandwidths up to 100 MHz.

Table 1.2 Cell user edge spectral efficiency

Test Environment	Downlink (bit/s/Hz)	Uplink (bit/s/Hz)
Indoor	0.1	0.07
Microcellular	0.075	0.05
Base coverage urban	0.06	0.03
High speed	0.04	0.015

1.3.4 Edge Spectral Efficiency

The (normalized) user throughput is defined as the average user throughput, i.e. the number of correctly received bits by users (the number of bits contained in the Service Data Unit (SDU) delivered to layer 3) over a certain period of time divided by the channel bandwidth and is measured in bit/s/Hz. The channel bandwidth for this purpose is defined as the effective bandwidth times the frequency reuse factor, where the effective bandwidth is the operating bandwidth normalized appropriately considering the UL/DL ratio. The cell edge user spectral efficiency is defined as 5% point of the Cumulative Distribution Function (CDF) of the normalized user throughput.

With X_i denoting the number of correctly received bits of user i, T_i the active session time for user i and ω the channel bandwidth, the (normalized) user throughput of user i, c_i, is defined according to (1.2). Table 1.2 lists the cell edge user spectral efficiency requirements for various test environments. The test environments are defined in the ITU-R Report M.2135

$$c_i = \frac{X_i}{T_i \cdot \omega} \tag{1.2}$$

1.3.5 Latency

The latency requirements are defined for the control plane (C-plane) and the user date plane (U-plane). The C-plane latency is typically measured as the transition time from different connection modes, e.g., from idle to active state. A transition time (excluding wireline network signalling delay) of less than **100 ms** shall be achievable from idle state to an active state in such a way that the user plane is established. The U-plane latency (also known as transport delay) is defined as the one-way transit time between an SDU packet being available at the IP layer in the MT/BS and the availability of this packet, Protocol Data Unit (PDU), at IP layer in the BS/MT. U-plane packet delay includes delay introduced by associated protocols and control

Table 1.3 Mobility classes for test environments

Test Environment	Mobility Classes Supported
Indoor	Stationary, Pedestrian
Micro Cellular	Stationary, Pedestrian, Vehicular (up to 30 km/h)
Base Urban	Stationary, Pedestrian, Vehicular
High speed	High speed vehicular, Vehicular

Table 1.4 Traffic channels, link data rates

	SE (bit/s/Hz)	Speed (km/h)
Indoor	1.0	10
Microcellular	0.75	30
Base coverage urban	0.55	120
High speed	0.25	350

signalling assuming the user terminal is in the active state. IMT-A systems shall be able to achieve a U-plane latency of less than **10 ms** in unloaded conditions (i.e., a single user with a single data stream) for small IP packets (e.g., 0 byte payload + IP header) for both DL and UL.

1.3.6 Mobility

The following mobility classes are defined:

- Stationary: $v = 0$ km/h
- Pedestrian: $0 \leq v \leq 10$ km/h
- Vehicular: $10 \leq v \leq 120$ km/h
- High speed vehicular: $120 \leq v \leq 350$ km/h.

For each test environment, a set of mobility classes shall be supported as reported in Table 1.3. For each mobility class, a minimum spectral efficiency is required as reported in Table 1.4. These values were defined assuming an antenna configuration of DL 4×2, UL 2×4.

1.3.7 Handovers

The handover interruption times are defined in Table 1.5 for intra- and inter-system handovers and for handovering within a spectrum band and among different spectrum bands.

Table 1.5 Handover interruption times

Handover type	Interruption time (ms)
Intra-frequency	27.5
Inter-frequency	
within a spectrum band	40
between spectrum bands	60

Table 1.6 VoIP capacity

Test Environment	Min VoIP capacity (active users/sector/MHz)
Indoor	50
Microcellular	40
Base coverage urban	40
High speed	30

1.3.8 VoIP Capacity

The Voice over IP (VoIP) capacity for the IMT-A candidates is derived assuming a 12.2 kbit/s codec with a 50% activity factor such that the percentage of users in outage is less than 2% where a user is defined to have experienced a voice outage if less than 98% of the VoIP packets have been delivered successfully to the user within a one way radio access delay bound of 50 ms. The specified codec is used for measurements and does not need to be effective part of the candidate system. The VoIP capacity is the minimum of the calculated capacity for either link direction divided by the effective bandwidth in the respective link direction. These values were defined assuming an antenna configuration of 4×2 in the DL and 2×4 in the UL. The antenna configuration is not a minimum requirement and the conditions for evaluation are described in Report ITU-R M.2135. The specifications indicate a minimum amount of VoIP connections to be supported, as in Table 1.6.

1.4 Mobile WiMAX

IEEE802.1-2009 aims to provide an upgrade for previous IEEE802.16 standard to fullfill the requirements of IMT-Advanced next generation mobile networks for the licesend spectrum bands. IEEE802.16-2009 provides continuing support for previous legacy 802.16d (WirelessMAN OFDMA) equipment. The working group is producing the following core documents. The System Requirements Document (SRD) [46] contains a set of possible de-

Table 1.7 Frequency bands for mobile WiMAX

Frequency bands
450–470 MHz
698–960 MHz
1710–2025 MHz
2110–2200 MHz
2300–2400 MHz
2500–2690 MHz
3400–3600 MHz

ployment scenarios and applications. The target performance and system features are specified.

The Evaluation Methodology Document (EMD) [44] proposes a complete set of parameters, models, and methodologies for the link-level and system-level evaluations. A set of spatial channel model parameters are specified to characterize particular features of MIMO radio channels.

The System Description Document (SDD) [45] reports the architecture and design of the 802.16 air interface amendment.

The working group will produce a final IMT-A Proposal to be submitted at the ITU-R.

The requirements for IEEE802.16 are divided in three groups: functional requirements, baseline requirements and a set of target requirements fulfilling the IMT-A specifications.

The general requirements are intended to address and supplement the requirements specified by the ITU-R for IMT-Advanced systems and are the minimum performance requirements for the MTs and BSs.

The functional requirements contains system level functional requirements targeting higher peak rates, lower latency, lower system overhead as well as PHY/MAC features enabling improved service security, QoS and radio resource management (RRM) for a system comprised solely of IEEE802.16 MTs and BSs.

The baseline requirements describe the performance requirements for IEEE802.16 systems specified in terms of absolute performance and relative performance with respect to that of the WirelessMAN-OFDMA Reference System.

IEEE802.16 systems will operate in RF frequencies less than 6 GHz and be deployable in licensed spectrum allocated to the mobile and fixed broadband services and shall be able to operate in frequencies identified for IMT-A. The IEEE802.16 operative frequency bands are reported in Table 1.7.

The frequency bands have been identified for IMT and/or IMT-2000 by World Administrative Radio Conference of 1992 (WARC92), World Radio Conference of 2000 (WRC2000) and World Radio Conference of 2007 (WRC07) and are reported in Table 1.7 and are described in Recommendation ITU-R M.1036-3. New frequency bands have been identified in the WRC07 but further work is ongoing inside the framework of ITU-R. IEEE802.16 shall support scalable bandwidths from 5 to 40 MHz. This bandwidth may be supported by single or multiple RF carriers.

IEEE802.16 shall support both Time Division Duplex (TDD) and Frequency Division Duplex (FDD) operational modes. The FDD mode shall support both full-duplex and half-duplex MT operation. A BS supporting FDD mode shall be able to simultaneously support half duplex and full duplex terminals operating on the same RF carrier.

IEEE802.16 shall support both unpaired and paired frequency allocations, with fixed duplexing frequency separations when operating in FDD mode. In TDD mode, the DL/UL ratio should be adjustable with the extreme of supporting DL-only configurations on a given carrier. In FDD mode, the UL and DL channel bandwidths may be different and should be configurable.

The IEEE802.16 standard defines minimum antenna requirements for the BS and MT. For the BS, a minimum of two transmit and two receive antennas shall be supported. For the MT, a minimum of one transmit and two received antennas shall be supported. This minimum is consistent with a 2×2 DL configuration and a 1×2 uplink configuration. IEEE802.16 shall support MIMO, beamforming operation or other advanced antenna techniques. IEEE802.16 shall further support single-user and multi-user MIMO techniques.

IEEE802.16 peak spectral efficiency achievable between a BS and an MT under ideal conditions is the highest theoretical data rate (normalized by bandwidth), which is the received data bits assuming error-free conditions assignable to a single mobile station, when all available radio resources for the corresponding link direction are utilized (that is excluding radio resources that are used for physical layer synchronization, reference signals or pilots, guard bands and guard times). IEEE802.16 peak Spectral Efficiencys (SEs) is specified in Table 1.8 where the baseline and the target specifications are reported. Latency, handovers timing, VoIP capacity and the other specifications are set to the same level as in the ITU IMT-A base requirements.

Table 1.8 Mobile WiMAX normalized peak data rate

Requirement type	Link	MIMO	Peak rate (bit/s/Hz)
Baseline	Downlink	2×2	8.0
Baseline	Uplink	1×2	2.8
Target	Downlink	4×4	15.0
Target	Uplink	2×4	6.75

1.5 Structure of the Monograph

The structure of the following five chapters is briefly outlined in this section. Chapter 2 presents introductory material for those not familiar with MIMO technologies. Single User (SU) MIMO theoretical capacity is presented. While numerous information theoretic results are available for SU MIMO, research is currently ongoing for several open problems in MU MIMO. After initial enthusiasm during the first experiments more than one decade ago, MIMO has found several practical difficulties in achieving the theoretical promised gains on a single-radio link. CL design of multiple-antenna PHY together with MAC can be the key technical approach to push for better resource exploitation with no (or limited) CSI feedback. Also, the MU MIMO-ARQ system proposed later in Chapter 5 is briefly outlined here.

In Chapter 3, adaptive strategies at the PHY layer are analyzed. Dynamic MIMO scheme selection based on channel status is presented for SU radio-links. The result is a PHY layer with variable SE driven by the average SNR. At low SNR, diversity schemes are used for lower bit error rates. When the channel is in a good state, spatial multiplexing can be used, transmitting multiple data streams over the same time-frequency PHY layer resource. A 4×2 MIMO system is also studied, which use a Layered STBC (LSTBC) where both spatial multiplexing and diversity gains are simultaneously achieved.

Then, a dynamic selection of STBC/SFBC over OFDMA is presented. A performance loss is experienced in STBCs if the channel is fast-varying (not constant over the length of the code). The constancy of the channel is usually not verified in the case of high velocities of the MTs. Based on Wide Sense Stationary – Uncorrelated Scattering (WSSUS) channel model and on the measures of delay spread and Doppler shift, the proposed system can adapt the symbols allocation over adjacent subcarriers or adjacent OFDMA symbols. This mitigates the detriment effect of time-varying channel for STBC. A closed-form analysis for the Symbol Error Probability (SEP) is proposed, where the error rate is derived using the Moment Generating Function (MGF) of the instantaneous SNR at the output of the Alamouti receiver for non-static channels.

Chapter 4 analyzes two protocols at the MAC layer: namely the ARQ and the DSA. The length of ARQ retransmissions is optimized jointly with adaptive MCS at the PHY layer. This results in bandwidth savings with time-varying channels that are modeled including an SNR log-normal distribution (shadowing). The DSA is a three-way handshake protocol used to admit service-flows in the network. Signaling blocking probability and admission control blocking probability are derived in closed-form for a block fading channel. The protocol is evaluated with fast-varying channels, providing insights over the DSA parameters impact on service-flow activation for fast MTs.

After the previous single-layer analysis, in Chapter 5 a set of cross-layer designs is proposed for a MU MIMO network. The system employs open-loop MIMO schemes – Alamouti and Vertical BLAST (VBLAST) – together with ARQ. The MAC layer is aware of per-antenna ACK and/or NACK. Several retransmission strategies are proposed, in cases of both SU and MU MIMO. Without loss of generality, the system uses a typical RR MAC scheduler; more advanced schedulers can be readily integrated in the current design. The performance is compared with fixed MIMO and adaptive MIMO schemes based on upper-bounded BERs, as presented in Chapter 3. The MU MIMO-ARQ state machine is located at the BS. The MU MIMO-ARQ system does not require any CSI feedback, but it is still able to capture a degree of MU diversity via a random user selection. The absence of CSI feedback is an important characteristic of the proposed design, as CSI feedback in MU MIMO requires a considerable amount of data capacity in particular with high mobility MTs.

Chapter 6 presents the protocol stack of an IEEE802.16 network. The protocol stack is described first in its classical form. Then, the issues related to the implementation of CL MU MIMO-ARQ strategies are discussed, and an enriched PHY/MAC interface is discussed. Several schedulers are reviewed for an integration in the MU MIMO-ARQ protocol proposed in Chapter 5, finally resulting in a complete PHY/MAC system that exploits CL techniques and MIMO antenna technology.

Part I

Single-Layer Adaptive Techniques

2

The Promises of MIMO

In the run for increasing capacity, the wireless research community is experiencing a barrier that the powerful Shannon formulation marks as insuperable. Channel coding has recently achieved impressive results, almost reaching the Shannon maximum limit: in [13] a BER of 10^{-5} is reached at SNR $= -0.46$ dB.[1]

The rising traffic demand calls for lightweight implementation of coding techniques but, more remarkably, requires new dimensions for achieving higher data rates. Space for further improvement in channel capacity was offered by the use of multiple antenna: MIMO has sparked a vibrant research area over the past decade. The added dimension of signaling coming from the multiple antenna elements provided the basis for a further leap in performance. Nowadays, MIMO is a fundamental component included in all the standardization activities for the current and future wireless networks.

The first results with MIMO systems showed impressive capacity gains. The first MIMO testbed [30] showed that capacity could be improved linearly with the minimum number of antennas installed at the transmitting or receiving sides of the radio link. A SE of 32 bit/s/Hz was achieved compared to almost unity efficiency for a Single Input Single Output (SISO) link over the same channel.

MIMO can support not only capacity improvements but can also be crucial in several other directions; the energy efficiency can be leveraged with an improved Signal-to-Noise Ratio (SNR) coming from array gain, and larger areas can be covered with the same transmitting power. MIMO also mitigates channel fading, reducing the error rates compared to SISO with equal link budget. MIMO can be used to partition various users in the cell, thus reducing interference and improving the frequency reuse factor.

[1] The BER is reached as closely as 0.03 dB from Shannon capacity with a considerably simple form of turbo-coding.

Opposed to all these opportunities, numerous practical limitations have been discovered between the SU MIMO theoretical results and system implementations.

MU MIMO, i.e. the scheduling of multiple users sharing the spatial dimensions of the channel, provides additional new scenarios for improvement. While SU MIMO techniques are relatively well understood, new investigations are needed to fully understand the relations among MIMO and MU networks. The interaction among multiple MIMO links impacts PHY, Medium Access Control (MAC) and Networking (NET) layers. They need to be addressed with a wide, comprehensive approach. During the past few years, the information theory community has achieved important breakthroughs in the understanding of MU MIMO systems, and these systems will play a crucial role in achieving the performance requirements in terms of better spectral efficiency, improved coverage and decreased latency.

In the following part of this chapter, first, the main results regarding SU MIMO capacity are briefly presented. The gains obtainable when MIMO is used in such a way are referred to as *MIMO processing gains*; the added degrees of signaling are used uniquely for detection and processing of the radio signals. These operations are fully contained at the PHY layer; the upper layers cannot interact with the dynamics of MIMO transmissions. Then, the *MIMO protocol gain* concept is presented, where the layers are interacting to grab all the opportunities that arise in MU MIMO networks.

2.1 Capacity in MIMO Channels

The capacity improvements promised by MIMO technology have been difficult to achieve. The main reasons of this implosion for the expected throughput gain come from:

- the difficulties of having an accurate and updated knowledge of the CSI at the transmitter side
- the propagation characteristics of the radio channel when low scattering is encountered
- the correlation among antenna elements, mainly at the MT.

SU MIMO channel capacity has been derived when perfect knowledge of CSI is assumed at both transmitter and receiver sides [35]. While this is usually considered a strong assumption regarding CSIT, there are cases where a very accurate knowledge of CSI at the Receiver (CSIR) is also impracticable, especially when MIMO schemes are used with very fast MTs.

The problem gets worse as the number of antennas grows: a high number of antennas is desired (despite high terminal cost), since, under favourable conditions, the capacity gain is increasing linearly with the minimum number of antennas available at the transmitter and receiver sides. A high number of antenna can exacerbate the problem of a performing CSI feedback and requires high-complexity receivers.

Some capacity results have been found for SU MIMO when only some statistical information about the channel is known, Channel Distribution Information (CDI), but several problems are still open in the field of MU MIMO for CDI-based feedbacks.

Despite these problems, even if CSI is perfectly known, correlation between antennas and poor scattering in the radio propagation environment can independently cause a big loss on the channel capacity.

2.1.1 MIMO Channel Theoretical Capacity

The Shannon capacity is defined as the maximum mutual information between the channel input and the channel output. The mutual information is the maximum data rate that the channel can sustain with an arbitrarily small error probability.

In the case of static channel, the previous definition is complete. When the channel is time-varying, several capacity definitions can be expressed since the capacity has a statistical distribution. The ergodic (Shannon) capacity is the maximum mutual information averaged over all the channel states.[2]

In Figure 2.1, the ergodic capacity for a SISO and several MIMO systems is reported with perfect CSIT and CSIR knowledge and perfectly uncorrelated antenna elements. The capacity improvement grows linearly with $\min(N_R, N_T)$.

The DL and UL CSI are represented by a $N_T \times N_R$ matrix $\mathbf{H}$ where $\mathbf{H} = \{h_{ij}\} \in \mathbb{C}$ with complex entries. Each entry represents the channel baseband equivalent h_{ij} from transmitting antenna jth to the receiving antenna ith. In the case where the channel reciprocity does not hold, DL and UL links can have substantially different capacities.

The channel capacity (which is deterministic, since the channel is static) is expressed as:

$$C = \max_{\mathbf{R}; Tr(\mathbf{R}) \leq P_T} \log |\mathbf{I}_{N_R} + \mathbf{HRH}^{\dagger}| \tag{2.1}$$

---

[2] Other common capacity definitions are: the outage capacity and the minimum-rate capacity.

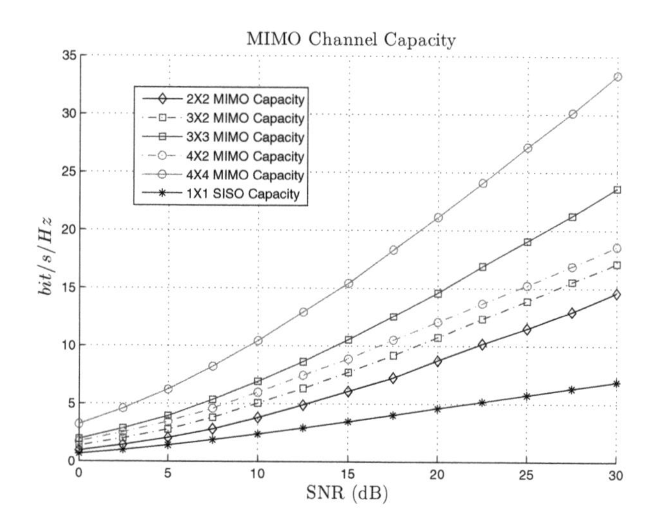

Figure 2.1 MIMO capacity with perfectly known CSIT and CSIR

where $\mathbf{I}$ is the identity matrix, $Tr()$ is the trace operator, $\dagger$ is the transpose conjugate operator, P_T is the total maximum transmit power and $\mathbf{H}$ is the MIMO channel matrix (CSI). The capacity depends on the covariance matrix of the input symbols: $\mathbf{R} = E[xx^\dagger]$, where x is the $N_T \times 1$ symbol vector at the input of the transmitting antenna system.[3] The maximum capacity is achieved when the inputs are Gaussian independent streams.[4]

In the case of fading (not constant channel), the result when CSIT and CSIR are perfectly known is straightforward, and it is expressed as the expectation of the capacity over all the channel instantaneous realizations:

$$C = E_{\mathbf{H}}\left[\max_{\mathbf{R};Tr(\mathbf{R})\leq P_T} \log|\mathbf{I}_{N_R} + \mathbf{HRH}^\dagger|\right] \tag{2.2}$$

When perfect CSIR is available, but CSIT is not available, the capacity has been evaluated by Foschini [31] and Telatar [29] assuming that the channel is Zero-Mean Spatially White (ZMSW); i.e., MIMO matrix $\mathbf{H}$ is assumed to have uncorrelated i.i.d. complex Gaussian entries. This assumption is very

[3] The input covariance matrix includes eventual precoding at the transmitter and antenna power allocation.

[4] In practice, the input streams cannot be Gaussian variables, but source data is modulated, i.e. represented with discrete symbols taken from digital modulations. This is also a source of capacity degradation for MIMO channels when low-order modulations are used.

strong since it does not consider the following effects usually encountered in practical systems:

- Antenna elements correlation: results in a matrix **H** with low rank and preferred propagation paths.
- Coloured Spatial Channel: radio channel with low scattering which results in a matrix **H** with low rank.

In the ZMSW model without CSIT, the problem is still to find the correct input covariance matrix to maximize the channel capacity; i.e. to choose **R** such that the following is maximum:

$$C(\mathbf{R}) = \max_{\mathbf{R};Tr(\mathbf{R} \leq P_T)} C(\mathbf{R}) \tag{2.3}$$

where $C(\mathbf{R})$ is

$$C(\mathbf{R}) = E_{\mathbf{H}}\left[\log |\mathbf{I}_{N_R} + \mathbf{HRH}^{\dagger}|\right] \tag{2.4}$$

In this case, it is found that the input covariance matrix that maximize the capacity is the scaled identity matrix; i.e., the power is equally divided among all the eigenmodes[5] and the ergodic capacity results as

$$C(\mathbf{R}) = E_{\mathbf{H}}\left[\log \left|\mathbf{I}_{N_R} + \frac{P_T}{N_T}\mathbf{HH}^{\dagger}\right|\right] \tag{2.5}$$

Foschini and Gans proposed a scalar code to achieve the predicted $C(\mathbf{R})$ capacity: the VBLAST vector coding scheme [30]. At 1% outage, at SNR = 12 dB, they showed, with a 12 × 12 antenna system, a spectral efficiency of 32 bit/s/Hz compared to 1 bit/s/Hz for the SISO channel with perfectly uncorrelated antennas and flat-fading channels.[6]

When also CSIR is not available, which is the case for highly MTs, the theoretical capacity achievable is even lower. *In the case that only CDIT and CDIR are known, the MIMO channel capacity has been demonstrated to grow only with the double log of the SNR with the number of antennas as a constant additive term [133].*

---

[5] Intuitively, since no information is known at the transmitter, the best uninformed choice is to divide the power equally over each antenna.

[6] In a more realistic system, as Gans reported in [33], the real capacity improvements were in the order of the 30% of the idealized setting. However, in [33], the results were still measured with independent antenna elements, totally avoiding any kind of interference.

2.1.2 MIMO Channel Rank

The problem of CSIT and CSIR availability has been briefly discussed; however, even if CSIT and CSIR are perfectly known, the propagation characteristics of the MIMO channel can cause severe losses in capacity. In particular, this depends on the MIMO channel matrix rank. High ranks indicate a radio channel with rich scattering, which leads to several independent spatial channels. If the channel rank is low, this means that only a principal propagation direction is available and that the spatial paths are highly correlated. MIMO capacity can be analyzed from the perspective of the Single Value Decomposition (SVD) of the channel matrix in order to highlight the impact of MIMO channel rank on the achievable MIMO capacity.

In 1999, Telatar showed in [29] that the MIMO channel can be converted in a set of parallel and independent SISO channel via an SVD of the MIMO matrix.[7]

The SVD theorem states that for any complex $N_R \times N_T$ matrix $\mathbf{H}$ with rank M, there exists an SVD of the form

$$\mathbf{H} = U\Lambda V^{\dagger} \tag{2.6}$$

where $U_{(N_R \times N_R)}$ and $V_{(N_T \times N_T)}$ are unitary matrices. The $N_R \times N_T$ matrix $\Lambda = [\lambda_{ij}]$ has $\lambda_{ij} = 0$ for all $i \neq j$, and $\lambda_{11} \geq \lambda_{22} \geq \lambda_{33} \geq \lambda_{44} \geq \lambda_{MM} \neq 0$ and $\lambda_{M+1,M+1} = \ldots = \lambda_{aa} = 0$, where $a = \min(N_T; N_R)$.

The singular values of $\mathbf{H}$, $\{\lambda_{ii};\ i = 1, 2, \ldots, M\}$, are the non-zero eigenvalues of $\mathbf{HH}^{\dagger}$, and hence are uniquely determined.

Let $\{\lambda_{mm}\}$ be the M non-zero singular values of $\mathbf{H}$. Then, $\left\{s_m = \lambda_{mm}^2\right\}$ are the M non-zero eigenvalues of either $\mathbf{HH}^{\dagger}$ or $\mathbf{H}^{\dagger}\mathbf{H}$. Since the number of transmitting or receiving antennas could differ, depending on the case where is $N_R > N_T$ or $N_R < N_T$, eigenvalues can be expressed in two ways. Following Telatar, a random matrix $\mathbf{W}$ can be defined as

$$\mathbf{W} = \begin{cases} \mathbf{HH}^* & \text{if } N_R < N_T \\ \mathbf{H}^*\mathbf{H} & \text{if } N_R \geq N_T \end{cases} \tag{2.7}$$

$\mathbf{W}$ is an $M \times M$ matrix, where $M = \min(N_R, N_T)$ and it is a *non-negative definite* matrix with real non-negative eigenvalues λ_i^2.

Each one of the set of parallel and independent SISO channel in which the MIMO channel can be decomposed, has the gain of one of the eigenvalues of matrix $\mathbf{W}$. The optimum power allocation that maximizes the channel

---

[7] SVD has several applications, e.g. in the field of matrix norm, rank and condition number calculations and, for linear system solving, noisy signal filtering and time series analysis.

capacity is the water-filling allocation: the power to be transmitted over each subchannel is given by

$$P_i = \left(\mu - \frac{1}{\lambda_i}\right)^+ \tag{2.8}$$

with $1 \leq i \leq \min(N_T, N_R)$, μ is the waterfill level, P_i is the power in the ith eigenmode and the $(x)^+$ operator is defined as $\max(0, x)$. The total capacity is thus the sum of the capacities over each eigenchannel:

$$C_T = \sum_{i=1}^{\min(N_T,N_R)} \log(\mu\lambda_i^2)^+ \tag{2.9}$$

Matrix **U** and matrix **V** can be used as pre-coding and post-coding unitary filters, respectively at the transmitter and receiver, in order to equalize the MIMO channel. SVD calculation is not straightforward to be implemented since it has an high computation complexity, with an order of O^3, and needs perfect CSIR and CSIT. Several advantages can be devised in SVD-based MIMO schemes: there is no cross-talk between spatial channels (eigenmodes) and conventional channel codes are readily used. SVD can be applied to channel matrixes of any size and any rank, thus SVD can be applied seamlessly to spatially correlated channels. Finally, SVD is optimal in the information theoretical sense, since unitary filters preserve covariance of the input data vector.

2.1.2.1 MIMO Channel Rank

The most important metric for a MIMO channel is the rank of **W**. The rank of **W** is directly related to the number of possible independent parallel subchannels that SVD decomposition can create, i.e. the spatial independent paths that are available in the MIMO channel. The worst case is when a unique path is present between the transmitter and the receiver: $\text{rank}(\mathbf{W}) = 1$. When the rank of **W** is maximum, the MIMO channel is said to be orthogonal and is

$$\text{rank}(W) = \min(N_R, N_T) \tag{2.10}$$

The rank corresponds to the maximum diversity gain d as defined in Equation (3.3); the orthogonal channel represents the capacity upper limit and the maximum diversity gain achievable.

High rank MIMO channel occurs when there is a rich scattering channel with little correlation between spatial channel paths. When the rank of matrix **W** is high, the capacity can be approximated showing two terms related to transmit array gain and SNR improvement due to the receiver array. The capacity can be expressed as

$$C_{\text{high}} \approx \underbrace{\min\left(N_T, N_R\right)}_{\text{array capacity advantage}} \cdot \underbrace{\log\left(1 + \frac{\frac{E}{2\sigma^2} N_R}{N_T}\right)}_{\text{receiver antenna SNR advantage}} \tag{2.11}$$

Low rank MIMO channel occurs when there is high correlation between the spatial subchannels. In the case of a MIMO channel with only one *non-zero* eigenvalue, the system is equivalent to a single antenna system with the same total power divided among N_T elements. A completely correlated $\mathbf{H}$ matrix ($\mathbf{H} \in M_{N_R \times N_T}$) can be represented as

$$\mathbf{H} = \begin{pmatrix} 1 & \ldots & 1 \\ \vdots & \ddots & \vdots \\ 1 & \ldots & 1 \end{pmatrix} \tag{2.12}$$

and correspond to the case of no diversity gain. The Spatial Division Multiple Access (SDMA) gain is null. The unique *non-zero* eigenvalue is given by

$$\lambda^2 = N_R \sum_{n=1}^{N_T} E\left\{h_{nn}^* h_{nn}\right\} = N_R N_T \tag{2.13}$$

where $h_{nn} = 0\ \forall n$ except for $n = 1$, i.e. $h_{11} \neq 0$. The capacity of the low-rank MIMO channel can be written as

$$\begin{aligned} C_{\text{low}} &= \sum_{n=1}^{\min(N_R, N_T)} \log\left(1 + \frac{\frac{E}{2\sigma^2} \lambda_n^2}{N_T}\right) \\ &\approx \log\left(1 + \frac{\frac{E}{2\sigma^2} \lambda^2}{N_T}\right) \end{aligned} \tag{2.14}$$

The channel behaves like a point to point channel with N_R times the received signal power achieved by Maximum Ratio Combining at the receiver.

The theoretical MIMO capacity gain presented assumes perfect CSI, both at the transmitter and receiver, and Spatially White Channels which are very difficult to achieve in practical system designs.

2.2 From MIMO Signal Processing Gain to MIMO Protocol Gain

Historically, MIMO has been employed as a PHY layer booster, where added degree of signaling is used to reduce error probability or to communicate at higher data rates.

For several reasons, MIMO has failed to meet all promises. This is mainly due to the difficulty in having decorrelated antenna elements (mainly at the MTs) and to the fact that the radio channel may present not-suitable conditions for spatial transmissions. The shift from SU to MU MIMO transmissions can be a key to recover some of performance losses with a CL design.

Cross-layer designs and layered-layerless architectures have been long debated in the research community, with evolutionary and revolutionary attitudes [89]. Without a doubt, layering plays an important role in the development of wireless standards and sustains commercially viable solutions, guaranteeing a stable setting for introducing technical ameliorations in a coordinated way. However, it is also widely admitted that layering imposes performance losses due to the sub-optimality of the current stack subdivision for multiple-access channels. In [88], an interesting overview of the layered-layerless topic is proposed, discussing the fundamental features of an MU communication network.

SU MIMO acts as a performance booster on the single radio link at the PHY layer only. Higher layers benefits from PHY greater SE but are not aware of what transmission is being used and cannot exploit multiple access mechanisms at the antenna level. The recent developments of CL techniques have made it clear that separating channel access, scheduling and MIMO schemes is limiting the total capacity attainable. MIMO is not regarded as a pure PHY technology anymore [34].

MU MIMO is based on the spatial sharing of the channel, where several users can be allocated along the spatial dimensions with an upgrade compared to SM schemes in SU links where all the spatial directions were allocated to the same user. MU MIMO requires an additional hardware cost and more complex scheduling and resource allocation techniques, but it can achieve very high SEs without bandwidth expansion. MU MIMO can provide gains in several directions:

- MU MIMO is more robust to channel rank-loss coming from low scattering and antenna elements correlation at the MT.

- MU MIMO can use SM schemes with single antenna MTs (via uplink collaborative schemes), reducing the cost of terminals and concentrating the wide antenna set at the BS.
- MU MIMO offers a capacity improvement exploiting multiuser diversity.

On the opposite side, some drawbacks limit the practicality of MU MIMO schemes. From a theoretical point of view, in MU MIMO DL Broadcast Channel the sum-rate capacity is achieved with Dirty Paper Coding (DPC) [19], which is an interference cancellation technique combined with user scheduling and power control.

First, the complexity of signal processing and the feedback needed for DPC are too heavy to be implemented in a practical system, and a plethora of simpler solutions have been proposed to approach, as closely as possible, the capacity upper-bound represented by DPC.

Second, user selection is required to solve an optimalization problem with a search space proportional to the number of users. A brute force search can be unfeasible for a big users set, and several trade-off solutions have been proposed, as in [123].

Third, the CSI feedback overhead can deeply affect the performance of the network in terms of throughput. A lower feedback MU MIMO scheduling with linear and non-linear precoding is presented in [34]. CSI feedback is challenging for two main reasons: feedback channel capacity is proportional to the number of users and antennas and, for highly mobile MTs, the information quickly becomes outdated. The heavy feedback poses serious limitations on the applicability of DPC and other closed-loop MIMO techniques even for slowly varying channels. For these reasons, partial (or no) feedback schemes have been proposed in [21,54,101,125], just to cite some recent works.

A simple MU MIMO system design with low feedback was proposed in [116]: Opportunistic Random BeamForming (ORBF). The system proposes a reduced feedback system assuming that the channel fading is not too fast and the number of users in the system is not too low. The antenna system is a "dumb" set of isotropic radiators where each one transmits a signal multiplied with a time-varying random complex coefficient (one or each antenna) communicated by the MTs to the BS. The aim is to artificially create large and fast channel fluctuations. Large fluctuations make it possible to schedule each user on the peak of its induced channel state and to maximize the system throughput. Fast-changing random coefficients are able to follow fast fading peaks but require high feedback rates; to maintain the amount of feedback tolerable, the coefficients cannot be chosen arbitrarily fast. However, if the period of the coefficients is longer than the correlation time of the fast fading process, ORBF will result in poor/no gain.

ORBF needs also a high number of users in order to have an effective system throughput gain. In [57], the issue of a low/realistic number of MTs has been addressed using more random coefficients inside a single transmission slot but exacerbating the rate of the feedback. In [59], a memory-based approach is proposed to exploit the time correlation of the channels when the number of users is low.

So far, the MIMO schemes presented need CSIT; the feedback problem can be solved by resorting to open-loop MIMO schemes, such as STBC or SM, that do not require CSIT. These schemes can be exploited to achieve diversity and/or rate gain; the diversity/spatial multiplexing trade-off[8] was settled down in [134]. The optimal scheme can be dynamically selected based on the spatial characteristics of the MIMO channel.

A switching system based on the Demmel condition number[9] of the channel matrix was presented in [41]. The system in [41] is able to select among MIMO schemes with equal SE: a 16-QAM Alamouti and a 4-QAM SM scheme, for instance, can be optimally chosen based on constellation distances expected at the receiver, as described in Section 3.5.1. The decision is made at the receiver based on locally estimated CSI and needs to be communicated back to the BS. However, adaptation to the channel state can be challenging in very fast channels and, as a general rule, the time to elapse the adaptive procedure needs to be, at maximum, one-tenth of the channel coherence time.

A variable spectral efficiency at the PHY layer can be obtained when adaptive MIMO schemes are chosen based on CSI maintaining the BER under a target BER threshold, as in [74] and as reported in Section 3.5.

2.3 Proposed MU MIMO-ARQ Protocol

A new approach for the exploitation of MIMO advantages is proposed in the second part of this work, which is based on the integration of MIMO, ARQ

[8] The formal definitions of diversity and rate gain are discussed in Sections 3.2 and 3.5.

[9] The Demmel Condition Number is a particular form of the condition number. The condition number, in general, expresses how reversible a matrix is. The condition number is related to the product of the norm of the matrix times the norm of the inverse. The Demmel condition number is defined in Section 3.5.1.

and scheduling without CSI feedback. With no loss of generality, a 2×2 antenna system is studied for DL transmission in an MU network.

A simpler MIMO-ARQ protocol was originally proposed in [23] for SU transmission. The proposed retransmission strategy is summarized here. The MT had a Minimum Mean Square Error (MMSE) with Successive Interference Cancellation (SIC) and Packet Combining (PC). The protocol for SU transmission is this. In the first time slot, two packets are sent in SM, then the system evolves depending on the Acknowledgements (ACKs) of the previously transmitted packets:

- When there are two packets in error, the packets are retransmitted with STBC (Alamouti).
- When there is only one packet in error, the packet is retransmitted in SM with a new packet. The packet is retransmitted from the weaker antenna such that the combined signal does not get excessive SNR. The weaker antenna is the one that gives the lowest post-processing SNR at the output of the receiver.
- When the two packets are received correctly, two new packets are sent to the user in SM.

As analyzed in detail in Chapter 5, the system proposed in [23] offers improvements over the adaptive MIMO systems presented in Chapter 3. However, the antenna allocation needs one bit feedback and the ARQ state machine cannot be realized as a free-running state machine.[10]

SU MIMO-ARQ could be straightforwardly applied in MU systems in a Time Division Multiplexing (TDM) fashion. When the user is scheduled, the system retransmits the packets until they are received or until the maximum number of ARQ retransmissions is achieved prior to moving to the next user in the scheduler queue.

With the SU strategy applied in TDM, in the case of dishomogeneous channel states, an MT in a bad channel state forces the system to wait until the ARQ protocol elapses the full retransmission time, reducing the total system throughput and serving the users with higher average delays.

A MU MIMO-ARQ protocol is proposed, which integrates the ARQ process and open-loop MIMO schemes in an MU network with an implicit Round Robin (RR) scheduler with any need of CSI feedback.

Several retransmission strategies are proposed. First, two strategies are studied. Assume that in the first time slot two packets are transmitted in SM.

[10] I.e., an ARQ state machine that does not need any feedback signal with the exception of the ACKs/NACKs at the MAC layer.

In this case, one strategy is to retransmit the wrong packet from both antennas to the same user; a second strategy is to retransmit the wrong packet with double power from a single antenna while the other is switched off.

A third strategy is proposed, which shows the best performance and observes the following: when there is one packet in error for the user (1), the packet is assumed to have been transmitted from the weaker antenna. The antenna allocation is thus switched (with no need of feedback signals), and the packet is retransmitted from the stronger antenna. A packet for a new user $(u + 1)$ is transmitted on the other antenna. The retransmission on the probably stronger antenna results in the lowest service time for the user (1), thus reducing the average delay in the system and maximizing the throughput. The antenna allocation follows the opposite principle with respect to Carvalho [23].

The new user is chosen randomly. The Hybrid SM transmission – i.e. the condition when the data streams are sent to different users – attains higher throughput by exploiting the multi-user diversity. The performance is compared to the case where the best user is chosen in order to maximize the throughput (i.e., the user with the highest post-processing SNR at the output of the receiver).

The proposed MU MIMO-ARQ protocol principal features are:

- No feedback at the BS for antenna allocation.
- No feedback at the BS for user selection.
- Open-loop STBC and SM are used (no CSIT needed) with seamless diversity/rate gain selection.
- Integration of the ARQ state machine with the packet scheduler.

An optimized SIC-MMSE receiver is proposed in Section 5.6. The MT selected for Hybrid SM transmission stores and exploits unintended packets sent in previous time slots to other users. The MT removes the contribution of the "overheard" packets to maximize the probability of the good reception of this packet.

The system design allows for integration of any kind of scheduler accounting for Quality of Service (QoS) attributes of the traffic flows. More advanced schedulers that can be combined inside the MIMO-ARQ system will be discussed in Chapter 6.

3

PHY Layer: MIMO Processing Gain

3.1 Introduction

In this chapter, MIMO is analyzed as a performance enhancer for SU PHY layer; the gain obtained at the PHY layer is referred to as the *MIMO processing gain*. When MIMO is employed at such layer, there is a performance enhancement which is originated from the processing of the signals with the added degrees of freedom coming from multiple antennas.

The gains of multiple antennas can be divided into three main categories. MIMO can provide *diversity* gain exploiting the independent fading in the multiple antenna links to transmit several coded versions of a data stream to enhance signal diversity, e.g. with STBCs and Space-Time Trellis Codes (STTCs). The first category is based on orthogonal designs [91, 106] while STTC distributes a trellis code over the antennas.

MIMO can also increase the rate (especially at high SNR) via *spatial* gain, with Spatial Multiplexing schemes. In these schemes, multiple data streams are transmitted from different antennas and arrives at the receiver with sufficient different spatial signatures. SM can be implemented with or without CSIT.

A third way is to use MIMO for beamforming or precoding. In closed-loop, the MT sends back the CSI information at the transmitter which "precodes" the signal, for instance, in order to maximize the Signal to Interference Noise Ratio (SINR) at the MT. The precoding is done via multiplication of each antenna signal with a complex coefficient, thus modulating the amplitude and the phase of the antenna signals individually such that when added together at the MT, they add constructively.

In general, the diversity gain can be obtained exploiting every domain of the wireless transmission. Typical examples are delay diversity and frequency diversity where signals are repeated over time or over different frequencies (see [90, chapter 14, p. 777]).

For multiple antennas, an effective way of providing diversity is through STBC; usually multiple transmissions are done over the two dimensions of space (antennas) and time (successive time slots), thus the name Space-Time Block Coding. STBCs assume that the channel is constant over the time of an STBC block. The performance is degraded if the channel is varying over the successive time slots.[1] The simplest STBC is the Alamouti code [5]. In its base version, it employs two antennas at the BS and one antenna at the MT. The scheme transmits two symbols – one from each antenna – during two successive time slots; thus, a total of four symbols per Alamouti-block are transmitted. Alamouti has been widely adopted because it is simple and requires low signal processing load at the receiver. Low power requirement is important because the MT is typically an energy-limited portable device.

A simple receiver, with a linear combiner and a single antenna, can extract an order of diversity equal to 2 if the assumption of constant channel over two time slots is met. If two receiving antennas are used at the MT, the diversity gain is four. No spatial gain is achieved with STBCs.[2]

In a Multi Carrier system, such as OFDM, Block Coding is also possible over the two domains of antenna and frequency with Space-Frequency Block Coding; SFBC coded blocks are transmitted during the same transmission slot but loaded over different subcarriers and radiated from different antennas.

Higher Spectral Efficiency can be achieved with SM schemes which transmit multiple data streams in the same time slot over the same frequency allocation resource. SM needs independently fading spatial channels. This condition is easily achievable in channels with a lot of scatterers and several spatial paths between the transmitter and the receiver. SM MIMO schemes support a data capacity that is related to the MIMO channel matrix rank. In particular, channel matrix with low condition numbers have highly correlated spatial paths, and the throughput achieved by SM is low.

At the receiver, a BLAST [30] or other Ordered Successive Interference Cancellation (OSIC) algorithms are used to extract the spatially multiplexed data streams from the received signal. Pure SM schemes give no diversity gain but a capacity gain depending on the number of parallel streams simultaneously transmitted. The number of antennas at the receiver has to be major or equal to the number of multiplexed streams.

---

[1] Such situation is analyzed in Section 3.6 for Alamouti scheme over uncorrelated Rayleigh channels.

[2] Spatial and diversity gains are formally defined in Sections 3.2 and 3.3, respectively.

The choice of MIMO schemes must be traded-off between total diversity STBC (robustness against fading) and SM (high rate). In [113], a rigorous definition of diversity rate is given and the tradeoff between diversity and spatial multiplexing is assessed. The impact on capacity is analyzed from an information theory point of view. The tradeoff on the usage of the antenna system gives the opportunity to choose adaptively several MIMO schemes. For instance, based on the instantaneous SNR, the best MIMO scheme (STBC or SM) can be chosen dynamically to maximize the capacity while maintaining the BER lower than a design threshold as presented in Section 3.5.

Hybrid schemes have been presented as an alternative to MIMO switching for achieving diversity and spatial multiplexing tradeoff; they combine together diversity and spatial multiplexing gain in the same transmission. In the case of four transmit antennas and two receiving antennas, it is possible to transmit two parallel Alamouti blocks, getting a spatial multiplexing of 2 and diversity gain of 2. The scheme is referred to as LSTBC and it is described in Section 3.3.1.

Summarizing, this chapter will analyzed the following MIMO schemes: Alamouti code with one receiving antenna (2×1) and with two receiving antennas (2×2), VBLAST (2×2) and LSTBC (4×2). The reference system is WiMAX IEEE802.16 in OFDM/OFDMA configuration. In Section 3.4, the MIMO support in the IEEE802.16 standard is described. The Adaptive Modulation and Coding (AMC) support in 802.16 is also presented, as the modulation order will be dynamically tweaked jointly with the MIMO schemes mentioned before.

Two solutions for MIMO adaptivity at the PHY layer are proposed. In Section 3.5, a dynamic PHY layer – switching among Alamouti, LSTBC and SM – is studied for having the best Spectral Efficiency (SE) guaranteeing an upper-bounded BER. In Section 3.6, the allocation of STBC over the time-frequency grid of the OFDMA-PHY layer is analyzed. The impact of the frequency correlation among subcarriers and the time variability due to mobility of the MT is studied. A general criterion for choosing between Space-Time or Space-Frequency Block Codes is assessed under the WSSUS channel assumption. The closed form of the Symbol Error Probability (SEP) is derived for the Alamouti scheme based on the Laplace transform of the channel fading distribution.

3.2 Space-Time Block Codes

Diversity techniques are employed to counter-fight fading and are widely used in wireless systems. The basic idea behind this approach is to transmit more copies of a signal; this makes possible to be more resistant to *deep fading*: if the receiver is supplied with more than one replica of the signal, transmitted over independent fading channels, the probability that all the copies are in a deep fade is greatly reduced.

For example, let us assume d frequency-non-selective Rayleigh independent channels and Binary Phase Shifting Key (BPSK) modulation: the slope of the BER curve over SNR depends on the number of replicas sent over the channels [90]. When SNR is sufficiently large (say greater than 10 dB) the bit error probability P_e can be approximated as

$$P_e(\mathrm{SNR}) = \left(\frac{1}{4\,\mathrm{SNR}}\right)^d \binom{2d-1}{d} \tag{3.1}$$

and for single or double diversity as

$$\begin{aligned} P_e(\mathrm{SNR})|_{d=1} &= \frac{1}{4}\mathrm{SNR}^{-1} \\ P_e(\mathrm{SNR})|_{d=2} &= \frac{3}{16}\mathrm{SNR}^{-2} \end{aligned} \tag{3.2}$$

The diversity gain (or order) is related to the number of parallel channels used. Since at high SNR the BER curve is characterized by its exponent, we can define the *diversity gain d* for a transmission scheme as

$$d = -\lim_{\mathrm{SNR}\to+\infty} \frac{\log P_e(\mathrm{SNR})}{\log \mathrm{SNR}} \tag{3.3}$$

where P_e is the bit error probability at fixed average SNR as expressed in (3.1).

The hypothesis of independent channels is not easily attainable and can be strengthen with several techniques. A first method is to exploit *frequency diversity*. The same signal is transmitted over d frequency carriers with a separation greater than the coherence bandwidth of the channel. This frequency separation ensures statistical independence between signal replicas. A second method provides diversity transmitting in successive time slots: *time diversity*. The signals are transmitted with a time separation greater than the channel coherence time.

A third interesting approach provides diversity using wideband signals. If a signal with bandwidth greater than the channel coherence bandwidth (B_C) is sent over the channel, the receiver can resolve all the multipath components via a RAKE receiver. The receiver exploits all the multipath replicas to achieve diversity, see [16] and references therein. Other diversity techniques regard *polarization* and *angle-of-arrival* diversity.

When a MIMO antenna system is available, diversity can be achieved over the antenna domain, via STBCs or STTCs.

The term STBC was first reported in [106] when Tarokh et al. proposed a new scheme for transmission over wireless Rayleigh-fading channels using multiple transmit antennas. The work in [104–106] can be thought of as a generalization for an arbitrary number of transmitting antennas of the Alamouti code previously published in 1998 [5]. Alamouti code is the simplest STBC construction possible and it was proposed as a simple diversity technique to achieve double diversity at an MT equipped with a single antenna with a BS equipped with two antennas: Alamouti code is able to "transfer" the double diversity from the BS to the MT, maintaining the more complex antenna system at the BS, given the assumption that the channel is constant over two symbols periods.

The design of STBC in [106] was based on the intuition that orthogonal designs[3] could be used for the design of wireless transmission schemes; with an underlying orthogonal structure, STBCs possess full diversity order and a decoupled Maximum-Likelihood (ML) decoding algorithm that avoids the exponential complexity of the ML receivers.

Orthogonal designs have particular properties; according to Radon's results on real orthogonal designs, they only exist if they have dimension of $\{1, 2, 4, 8\}$. A real orthogonal design can be used for code construction with real symbols (for instance for Pulse Amplitude Modulation (PAM) or BPSK modulations). In [105] Radon's results have been extended to both rectangular and complex orthogonal designs (code construction for complex modulations).

An important characteristic of STBCs is the design rate: the rate of an STBC is defined as the ratio of the number of data symbols over the decoding delay of these symbols at the receiver (for any given number of transmit antennas).

An STBC design with a unitary rate ($R_{\text{stbc}} = 1$) can send, for instance, two symbols over two transmission time slots without loss in bandwidth. In

[3] Orthogonal designs have been studied in the field of pure mathematics by Radon in [91].

this case, the antenna resource is used to add diversity without sacrificing the capacity [106, corollary 3.3.1, p. 756].

For real orthogonal designs, it was shown that it exists a full-rate design for any given number of transmit antennas; STBCs can achieve the maximum possible rate for any given number of transmit antennas.

In the case of complex signal constellations, such as an arbitrary order Phase Shifting Key (PSK) – order different from two – and Quadrature Amplitude Modulation (QAM), the authors were not able to prove the existence of orthogonal designs for any number of transmit antennas, even if hypothesized. In [106], the complex orthogonal design with a dimension of 2 (Alamouti code) was pointed out as an example of rate-one complex design that was part of a probable bigger family of complex orthogonal designs with unitary rate. Contrary to this logical intuition, in [68], it was proved that there does not exist any rate-one complex designs except of that with dimension equal to two, even if an infinite decoding delay is tolerated at the receiver. This result is in sharp contrast to the existence of rate-one generalized real orthogonal designs for any given number of transmit antennas.

3.2.1 Alamouti Scheme

The Alamouti scheme [5], also referred to as Space-Time Transmit Diversity (STTD), needs two transmit antennas at the BS and provides a transmit diversity of 2 when the MT is equipped with a single antenna. In this configuration, the Alamouti scheme is able to provide a diversity gain of two using the double antenna system at the BS instead of requiring 2 antennas at the MT. This is very beneficial because the added costs of the multiple transceivers and antennas are concentrated in a central point and not replicated over each MT. The Alamouti scheme can offer a diversity gain of 4 in 2×2 MIMO system.

In the case of complex modulation, the Alamouti scheme is the only scheme for complex modulations known (and possible) to be fully orthogonal, with full diversity and rate $R_{\text{stbc}} = 1$. The Alamouti coding matrix can be expressed as

$$\mathbf{A} = \begin{pmatrix} s_1 & s_2 \\ -s_2^* & s_1^* \end{pmatrix} \tag{3.4}$$

where symbol $s = a_{ij}$ is transmitted at the ith time slot over jth transmit antenna. The receiver is a linear combiner [5] that reconstructs data symbols by using orthogonal properties of space-time coding matrix $\mathbf{A}$.

Channel variations, that occur during two symbol time intervals, are a severe source of performance degradation. In [5, 105], the channel is assumed to be constant over the time length of the code. However, this assumption is usually unverified, especially when the code length is quite long and MTs are moving at high speed. In this case, the channel value at time t can be expressed in relation to the preceding one $(t-1)$ with a correlation coefficient $\rho < 1$.

In Section 3.2.2, a closed-form SEP for the Alamouti scheme in a perfectly correlated channel is derived using the MGFs. In [115], the BER for the Alamouti scheme over non-perfectly correlated channels has been derived in closed-form for BPSK in a narrow band Rayleigh channel. In Section 3.6.4, the closed-form BER expression has been extended for a QAM modulation of any order and used as a benchmark for the performance evaluation of the dynamic allocation of STBC proposed in Section 3.6.5.

3.2.2 Closed-Form SEP Calculation in Correlated Channels

A calculation technique for a closed-form SEP is thoroughly analyzed for M-QAM modulation with a coherent receiver over a perfectly correlated Rayleigh fading channel. The results obtained herewith will be used later in Section 3.6.4 as a basis to obtain the SEP for Alamouti scheme over non-perfectly correlated channel. The technique is based on the alternative expression of the Q-function found by Craig in 1991 [20], and on the alternative expression of the squared Q-function as derived in [98].

3.2.2.1 SEP Calculation Using Moment Generating Function

Closed-form average SEP evaluation is usually difficult; the average SEP is obtained integrating the symbol error rate over the instantaneous SNR distribution. The conditional SEP is, in general, an expression containing non-linear functions of the SNR. The non-linear integrand depends on the modulation and coding scheme adopted, but even in the case of a simple BPSK, the error rate depends on the Q-function, which is an indefinite integral of an exponential function with no closed-form integral.[4] The MGF of the channel fading is the Laplace transform of the fading distribution. The

[4] The Q-function is defined as

$$Q(x) = 1/\sqrt{2\pi} \int_x^{\infty} \exp(-y^2/2)dy = 1/2 \operatorname{erfc}(x/\sqrt{2})$$

SEP integral can be evaluated more easily plugging the MGF of the SNR distribution for the following reasons:

- the integrand function is in the form of a rational function (instead of non-linear Q function),
- the integration interval is finite avoiding the calculation of the indefinite integral over all the instantaneous SNR distribution.

3.2.2.2 SEP for M-QAM Modulations in AWGN Channels

As is well known, the SEP for QAM modulation of order M is obtained as the quadrature combination of two Amplitude Modulations (AM) with half symbol energy and order $\sqrt{M}$:

$$\mathrm{SEP}(E_S)|_{M-\mathrm{QAM},E_S} = 1 - \left[1 - \mathrm{SEP}(E_S)|_{\sqrt{M}-\mathrm{AM},E_S/2}\right]^2 \tag{3.5}$$

Recalling the SEP for AM:

$$\mathrm{SEP}(E_S)|_{M-\mathrm{AM},E_S} = 2\left(\frac{M-1}{M}\right) Q\left(\sqrt{\frac{6E_S}{N_0(M^2-1)}}\right) \tag{3.6}$$

the M-QAM SEP for the AWGN channel can be expressed as

$$\begin{aligned}\mathrm{SEP}(E)|_{M-\mathrm{QAM},E_S} &= 4\left(\frac{\sqrt{M}-1}{\sqrt{M}}\right) Q\left(\sqrt{\frac{3E_S}{N_0(M-1)}}\right)\\ &\quad - 4\left(\frac{\sqrt{M}-1}{\sqrt{M}}\right)^2 Q^2\left(\sqrt{\frac{3E_S}{N_0(M-1)}}\right)\end{aligned} \tag{3.7}$$

For the calculation of the average SEP, the Q-function and Q-squared function are involved as integrands of an indefinite integral. An alternative Q-function expression is found by Craig in 1991 [20] via a variable transformation; it can be obtained extending the classical Q-function over two dimensions in polar coordinates and then integrating over half plane. The alternative form is written as

$$\begin{aligned}Q_{AL}(x) &= \frac{1}{\sqrt{2\pi}}\int_x^{\infty} \exp\left(-\frac{y^2}{2}\right) dy\\ &= \frac{1}{\pi}\int_0^{\pi/2} \exp\left(-\frac{x^2}{2\sin^2\theta}\right) d\theta\end{aligned} \tag{3.8}$$

For the Q-squared function, based on the previous derivation and following appendix 4A in [98], a similar alternative form can be obtained as the following:

$$
\begin{aligned}
Q^2_{\mathrm{AL}}(x) &= \frac{1}{\sqrt{2\pi}}^2 \int_x^{\infty} \exp\left(-\frac{y^2}{2}\right) dy \cdot \int_x^{\infty} \exp\left(-\frac{y^2}{2}\right) dy \\
&= \frac{1}{\pi} \int_0^{\pi/4} \exp\left(-\frac{x^2}{2\sin^2\theta}\right) d\theta
\end{aligned}
\tag{3.9}
$$

The Q-squared function integration can be obtained simply changing the integration interval in the expression of (3.8). Despite its apparent simplicity, however, this change of interval leads to closed-forms that can be far more complicated compared to $Q(x)$. We will see in the next section, that a simple closed-form is available when the channel fading has Rayleigh distribution.

Using the alternative functions reported in (3.8) and in (3.9), the SEP for AWGN channel can be readily calculated as

$$
\begin{aligned}
\mathrm{SEP}(E) &= \frac{4}{\pi}\left(\frac{\sqrt{M}-1}{\sqrt{M}}\right) \int_0^{\pi/2} \exp\left(-\frac{E_s}{N_0}\frac{3}{2(M-1)\sin^2\theta}\right) d\theta \\
&\quad - \frac{4}{\pi}\left(\frac{\sqrt{M}-1}{\sqrt{M}}\right)^2 \int_0^{\pi/4} \exp\left(-\frac{E_s}{N_0}\frac{3}{2(M-1)\sin^2\theta}\right) d\theta
\end{aligned}
$$

3.2.2.3 SEP with Generic Channel Fading

In the case of fading channel, the instantaneous SNR distribution needs to be integrated in order to obtain the average SEP. For the narrowband channel analyzed in Section 3.6, the received signal amplitude is multiplied by the channel fading amplitude which is a random variable β with a mean-square value of Ω. The Probability Density Function (PDF) of the fading amplitude ($p_\beta(\beta)$) is characteristic of the radio propagation. After being multiplied by the channel fading, the signal is disturbed with the additive noise at the receiver. The noise at the receiver is usually considered independent from the fading statistic and defined by a one-sided power spectral density of N_0 (W/Hz). Following the assumption of statistic independence among fading and added noise at the receiver, an instantaneous SNR can be defined as

$$
\gamma = \beta^2 \frac{E_b}{N_0} \tag{3.10}
$$

where E_b is the energy per bit. The PDF of the SNR can be derived with the variable transformation defined in (3.10) as

$$p_\gamma(\gamma) = \frac{p_\beta(\sqrt{\Omega\gamma/\overline{\gamma}})}{2\sqrt{\gamma\overline{\gamma}/\Omega}} \tag{3.11}$$

where the average SNR is $\overline{\gamma} = \Omega E_b/N_0$.

The conditional SEP for a certain instantaneous SNR, is obtained with the AWGN SEP where E_S/N_0 is replaced by $\gamma \log_2(M)$; $\log_2(M)$ is the bits per symbol which depends on the QAM modulation order. The average SEP is obtained from the conditional SEP integrated over all the real axis and the following has to be solved:

$$I = \int_0^\infty Q(a\sqrt{\gamma}) p_\gamma(\gamma) d\gamma \tag{3.12}$$

where a is a constant which depends on the modulation adopted.

Let us first define the integral expressions for Q and Q-squared functions over the fading distribution

$$\begin{aligned} I_Q(a,\gamma) &= \int_0^\infty Q(a\sqrt{\gamma}) p_\gamma(\gamma) d\gamma \\ &= \int_0^\infty \frac{1}{\pi} \int_0^{\pi/2} \exp\left(-\frac{a^2\gamma}{2\sin^2\theta}\right) d\theta \; p_\gamma(\gamma) d\gamma \\ &= \frac{1}{\pi} \int_0^{\pi/2} \left[\int_0^\infty \exp\left(-\frac{a^2\gamma}{2\sin^2\theta}\right) p_\gamma(\gamma) d\gamma \right] d\theta \end{aligned} \tag{3.13}$$

The MGF[5] of the SNR distribution is defined as

$$M_\gamma(s) = \int_0^\infty p_\gamma(\gamma) \exp(s\gamma) d\gamma \tag{3.14}$$

and is the Laplace transform of the SNR probability density. The MGF permits to easily evaluate the SEP for several channel statistics and modulation and coding schemes avoiding the complex indefinite integral with non-linear integrand.

[5] The name "Moment Generating Function" comes from the fact that the nth moment of the random variable $X(t)$ can be obtained simply deriving n times the MGF and evaluating for $t = 0$:

$$E\{X^n\} = \frac{d^n}{dt^n} M_X(t)|_{t=0}$$

Noting that

$$M_\gamma(s)\Big|_{s=-\frac{a^2}{2\sin^2\theta}} = \int_0^\infty \exp\left(-\frac{a^2\gamma}{2\sin^2\theta}\right) p_\gamma(\gamma)d\gamma \tag{3.15}$$

is the MGF of the instantaneous SNR distribution evaluated in $s = -a^2/2\sin^2\theta$, the averaged Q-function can be written as

$$I_Q(a,\gamma) = \frac{1}{\pi}\int_0^{\pi/2} M_\gamma\left(-\frac{a^2}{2\sin^2\theta}\right)d\theta \tag{3.16}$$

Using (3.9), the Q-squared function integrated over the SNR distribution $p_\gamma(\gamma)$ is

$$\begin{aligned} I_{Q^2}(a.\gamma) &= \int_0^\infty Q^2(a\sqrt{\gamma})p_\gamma(\gamma)d\gamma \\ &= \frac{1}{\pi}\int_0^{\pi/4} M_\gamma\left(-\frac{a^2}{2\sin^2\theta}\right)d\theta \end{aligned} \tag{3.17}$$

Recalling $\text{SEP}(E_S)|_{M-\text{QAM},E_S}$ from (3.7), substituting E_S/N_0 with $\gamma\log_2(M)$ and integrating over the fading distribution, yields

$$\text{SEP}(\overline{\gamma}) = 4\left(\frac{\sqrt{M}-1}{\sqrt{M}}\right)I_Q(a,\gamma) - 4\left(\frac{\sqrt{M}-1}{\sqrt{M}}\right)I_{Q^2}(a,\gamma) \tag{3.18}$$

where for M-QAM modulations

$$a^2 = 3\frac{\log_2(M)}{M-1}$$

3.2.2.4 Rayleigh Fading Channel Statistic

In this section, the previous expressions are calculated for the particular case of Rayleigh fading. The Rayleigh fading is used to model non-Line of Sight (LoS) channel with multipath fading. The fading statistic is expressed as

$$p_\beta(\beta) = \frac{2\beta}{\Omega}e^{-\beta^2/\Omega} \tag{3.19}$$

and is defined for $\beta \geq 0$. The instantaneous SNR probability density is thus

$$p_\gamma(\gamma) = \frac{1}{\overline{\gamma}}e^{-\gamma/\overline{\gamma}} \tag{3.20}$$

again defined for $\gamma \geq 0$. The MGF for the Rayleigh statistic is

$$M_{\gamma}^{\mathrm{Ray}}(s) = \frac{1}{1 - s\overline{\gamma}} \tag{3.21}$$

Other common fading models (such as Nakagami-m, Rice, Hoyt) and relative MGFs can be found in [98].

In this case, the following can be derived:

$$\begin{aligned} I_{Q}^{\mathrm{Ray}}(a, \overline{\gamma}) &= \frac{1}{\pi} \int_{0}^{\pi/2} M_{\gamma}^{\mathrm{Ray}}(s) d\theta \\ &= \frac{1}{\pi} \int_{0}^{\pi/2} \left(1 + \frac{a^2 \overline{\gamma}}{2 \sin^2 \theta}\right)^{-1} d\theta \\ &= \frac{1}{2} \left(1 - \sqrt{\frac{a^2 \overline{\gamma}/2}{1 + a^2 \overline{\gamma}/2}}\right) \end{aligned} \tag{3.22}$$

For the Q-squared function with Rayleigh fading, the following derivation holds:

$$\begin{aligned} I_{Q^2}^{\mathrm{Ray}}(a, \overline{\gamma}) &= \frac{1}{\pi} \int_{0}^{\pi/4} M_{\gamma}^{\mathrm{Ray}}(s) d\theta \\ &= \frac{1}{\pi} \int_{0}^{\pi/4} \left(1 + \frac{a^2 \overline{\gamma}}{2 \sin^2 \theta}\right)^{-1} d\theta \\ &= \frac{1}{4} \left[1 - \sqrt{\frac{a^2 \overline{\gamma}/2}{1 + a^2 \overline{\gamma}/2}} \left(\frac{4}{\pi} tan^{-1} \sqrt{\frac{1 + a^2 \overline{\gamma}/2}{a^2 \overline{\gamma}/2}}\right)\right] \end{aligned} \tag{3.23}$$

Finally, considering that

$$a^2 = 3 \frac{\log_2(M)}{M - 1} \quad \text{and} \quad \overline{\gamma_s} = \overline{\gamma} \log_2(M)$$

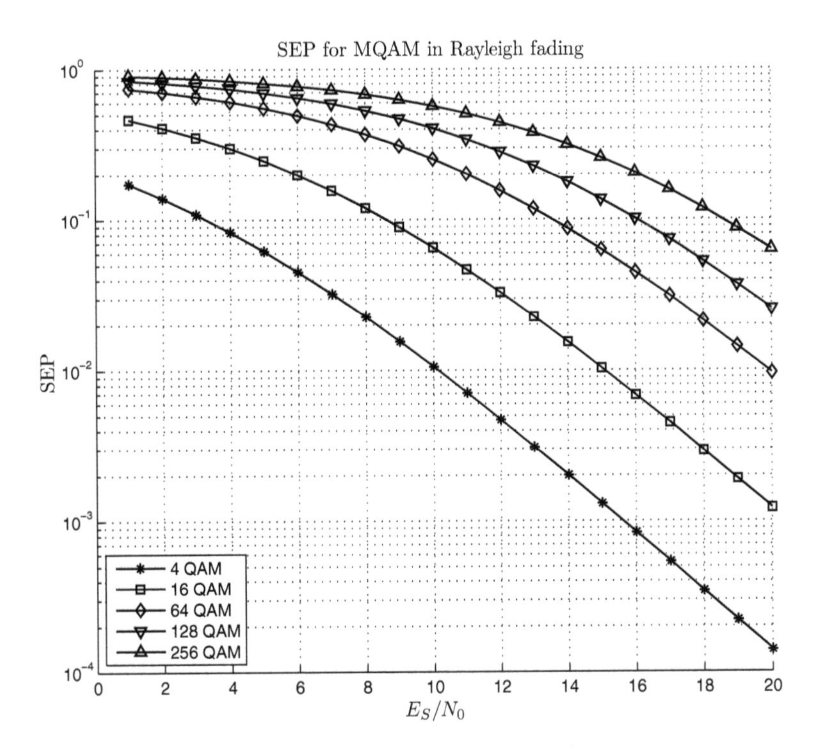

Figure 3.1 SEP for various order of Quadrature Amplitude Modulations

the SEP for Rayleigh fading with coherent QAM modulation, reported also in Figure 3.1, is obtained as

$$\begin{aligned}\mathrm{SEP}(\overline{\gamma}) &= 2\left(\frac{\sqrt{M}-1}{\sqrt{M}}\right)\left(1-\sqrt{\frac{3\overline{\gamma}\log_2(M)/2}{M-1+3\overline{\gamma}\log_2(M)/2}}\right)\\ &\quad-\left(\frac{\sqrt{M}-1}{\sqrt{M}}\right)^2\left[1-\sqrt{\frac{3\overline{\gamma}\log_2(M)/2}{M-1+3\overline{\gamma}\log_2(M)/2}}\right.\\ &\quad\left.\cdot\left(\frac{4}{\pi}\tan^{-1}\sqrt{\frac{M-1+3\overline{\gamma}\log_2(M)/2}{3\overline{\gamma}\log_2(M)/2}}\right)\right] \qquad (3.24)\end{aligned}$$

3.3 Spatial Multiplexing

The diversity gain has been defined in Section 3.2. The definition of multiplexing gain is similarly expressed as

$$r = \lim_{\text{SNR}\to+\infty} \frac{R(\text{SNR})}{\log(\text{SNR})} \tag{3.25}$$

where $R(\text{SNR})$ is the rate achievable for a specified SNR.

As already discussed in Chapter 2, the ergodic capacity of the MIMO channel in SM scales with the minimum of the number of receiving and transmitting antennas as

$$R(\text{SNR}) \approx m \in (N_T, N_R) \log(\text{SNR}) \quad \text{(bps/hz)} \tag{3.26}$$

The first and most famous SM implementation is the Vertical BLAST. Independent data streams are spatially multiplexed and each data stream is referred to as a layer. Other forms have been proposed in literature such as Diagonal BLAST (DBLAST) and in [52, 84, 130].

Since VBLAST coding is operating over a single transmission time slot, it is a space-only coding and the coding matrix is reduced to a vector (VBLAST is sometimes called Vector Modulation):

$$\mathbf{C} = \begin{pmatrix} s_1 \\ s_2 \end{pmatrix} \tag{3.27}$$

Layers can be separated at the receiver if the number of receiving antennas is $N_R > N_{\text{streams}}$. While Alamouti and its hybrid version (two layered Alamouti blocks) do not exploit all the degree of freedoms of the MIMO channel, VBLAST can extract the complete spatial gain. On the other hand, VBLAST does not achieve any diversity gain.

Optimum decoding order has been demonstrated in [122] that is obtained detecting the highest SNR layer at each OSIC step. At each stage, linear combinatorial nulling is applied to remove interference caused by undetected layers. VBLAST receiver based on Zero Forcing (ZF) linear detection technique can be briefly summarized with the following steps. The receiver procedure is composed by the following phases: initialization, cancellation and update.

Initialization:

I1 $\mathbf{G}_1 = \mathbf{H}^+$, the channel matrix is inverted

I2 $k_1 = \arg\min_j \|(\mathbf{G}_1)_j\|^2$, the index of the column of $\mathbf{G}_1$ with minimum squared norm is found

Then a loop is performed for cancellation – the number of times the loop is executed is equal to the number of receiving antennas N_R:

Cancellation:

C1 find the highest SNR received signal out of N_R signals available

C2 $\mathbf{w}_{k_i} = (\mathbf{G}_i)_{k_i}$, compute the weight vector $\mathbf{w}_{k_i}$ selecting the k_ith column of the inverted channel matrix $\mathbf{G}_i$

C3 perform signal detection (weighting and slicing)

$$y_{k_i} = \mathbf{w}_{k_i}^T \mathbf{r}_i$$
$$\hat{a}_{k_i} = Q(y_{k_i})$$

C4 operate interference cancellation

$$\mathbf{r}_{i+1} = \mathbf{r}_i - \hat{a}_{k_i}(\mathbf{H})_{k_i}$$

Update

U1 update channel matrix before next OSIC iteration

$$\mathbf{G}_{i+1} = \widetilde{\mathbf{H}}_{k_i}^+$$

U2 $k_{i+1} = \arg\min_{j \notin \{k_1, \ldots k_j\}} \|(\mathbf{G}_{i+1})_j\|^2$, the index for the next weighting vector is found

U3 $i + 1 \rightarrow i$, update of the loop index for the next iteration

where $Q()$ is the detection operator, $\mathbf{G}_i = \mathbf{H}^+$ is the channel matrix pseudo-inverse at the ith step of the detection algorithm, $(\mathbf{G}_i)_j$ is the jth row of $\mathbf{G}_i$, $\widetilde{\mathbf{H}}_{k_i}$ denotes the matrix obtained by zeroing columns $k_1, k_2, \ldots, k_i$ and $(\ )^+$ denotes the Moore–Penrose pseudo-inverse. The process is repeated until the last interference-free layer has been detected.

As in the former hybrid transmission scheme, the receiver here described can be used in a multiuser context and data layers transmitted with an SM approach can belong to a single or multiple users.[6]

[6] In 802.16 standards, the users to whom the streams belong to, are specified in the frame map broadcasted from the BS. This transmission is identified as Hybrid SM transmission. Hybrid SM will be studied in conjunction with ARQ. Hybrid SM will form the basis for the MU MIMO-ARQ protocol, proposed in Chapter 5, which aims at achieving better opportunistic scheduling and network capacity.

3.3.1 LSTBC

STBC from orthogonal designs, proposed in [106], has attracted considerable attention thanks to their easy ML decoding and full diversity. A drawback of such designs is that the orthogonality cannot be achieved with an high number of antennas without losing the unitary rate of the scheme. For increasing the symbol rates, the orthogonal STBC have been generalized by relaxing the orthogonality property of the design [50].

LSTBC is a Quasi-Orthogonal Space-Time Block Code (QOSTBC): the scheme is operating over a 4×2 MIMO system: two Alamouti blocks are "layered", i.e., transmitted from two set of two antennas each. Each Alamouti block is transmitted in Spatial Multiplexing.

In LSTBC, orthogonality is lost, but the full diversity gain and the receiver linearity are preserved. Of course, this scheme is a tradeoff between a full-diversity (Alamouti) and a SM approach that would be offering maximum capacity gain but no diversity gain, as shown in Section 3.5.

Since layered schemes are subject to interference among symbols transmitted from different antennas, a BER degradation is experienced with respect to the orthogonal case. The receiver needs a number of receiving antennas $N_R > N_{\text{blocks}}$, where N_{blocks} are the orthogonal blocks which have been spatially multiplexed (or layered). In the case of a 4-antenna BS, the STBC coding matrix can be expressed as

$$\mathbf{B} = \begin{pmatrix} s_1 & s_2 & s_5 & s_6 \\ -s_1^* & s_2^* & -s_6^* & s_5^* \\ s_3 & s_4 & s_7 & s_8 \\ -s_4^* & s_3^* & -s_8^* & s_7^* \end{pmatrix} \tag{3.28}$$

where rows represent the antennas and the columns the transmission time slots. Two Alamouti blocks are transmitted at the same time from a subset of the available BS antennas. This solution achieves the same transmit diversity as in the classic Alamouti scheme and a spatial gain that is upper limited to 2. A spatial gain equal to two is achieved when the channel matrix has full rank. In this case, no detection loss is experienced since cross-stream interference is null. However, if the MIMO channel rank is low, the receivers can have difficulties to extract the multiplex streams.

The receiver operates as follows [50]. Let us consider a system with a number of transmitting antennas $N_T = 4$ and a number of receiving antennas $N_R = 2$. The channel coefficient between the ith transmitting element and jth receiving one at the nth signalling time is $h_{ij(n)}$ and $\mathbf{s} = [s_1, s_2, s_3, s_4]^T$.

At the receiver, the following composition of the two $N_R \times 1$ received vectors $\mathbf{y}$ at time n and $n+1$ is considered:

$$\begin{bmatrix} \mathbf{y}_{(n)} \\ \mathbf{y}^*_{(n+1)} \end{bmatrix} = \begin{bmatrix} h_{11(n)} & h_{21(n)} & h_{31(n)} & h_{41(n)} \\ h_{12(n)} & h_{22(n)} & h_{32(n)} & h_{42(n)} \\ h^*_{21(n+1)} & -h^*_{11(n+1)} & h^*_{41(n+1)} & -h^*_{31(n+1)} \\ h^*_{22(n+1)} & -h^*_{12(n+1)} & h^*_{42(n+1)} & -h^*_{32(n+1)} \end{bmatrix} \mathbf{s} + \begin{bmatrix} \mathbf{n}_{(n)} \\ \mathbf{n}_{(n+1)} \end{bmatrix} \tag{3.29}$$

where $\mathbf{n}_{(n)}$ is the $N_R \times 1$ noise vector at time n.

Channel matrix $\mathbf{H}$ has peculiar orthogonality properties that can be exploited in order to separate the spatially multiplexed Alamouti blocks. Let us make a partition of the channel matrix as $\mathbf{H} = [\mathbf{H}_1|\mathbf{H}_2]$.

The Least-Square (LS) receiver is obtained from the Moore–Penrose pseudo-inverse $\mathbf{H}^+$, as follows:

$$\hat{\mathbf{s}} = \mathbf{H}^+\mathbf{y} = \mathbf{U}\mathbf{V}\,\mathbf{y} = \begin{bmatrix} \mathbf{H}_1^H\mathbf{H}1 & \mathbf{H}_1^H\mathbf{H}_2 \\ \mathbf{H}_2^H\mathbf{H}_1 & \mathbf{H}_2^H\mathbf{H}_2 \end{bmatrix}^{-1} \begin{bmatrix} \mathbf{H}_1^H \\ \mathbf{H}_2^H \end{bmatrix} \mathbf{y} \tag{3.30}$$

The matrix $\mathbf{U}$ contains the auto-products of Alamouti blocks $\mathbf{H_1}$ and $\mathbf{H_2}$ on the principal diagonal and it has two cross-products representing the interferences due the spatial multiplexing. Let $(\ )^H$ and $(\ )^T$ be the hermitian and real transpose operators, respectively. Due to Alamouti orthogonality, the auto-products of Alamouti blocks are

$$\mathbf{H}_1^H\mathbf{H}_1 = k_1\mathbf{I}, \qquad \mathbf{H_2}^H\mathbf{H_2} = k_2\mathbf{I}$$

where

$$k_1 = \|\mathbf{h}_1\|^2 + \|\mathbf{h}_2\|^2, \qquad k_2 = \|\mathbf{h}_3\|^2 + \|\mathbf{h}_4\|^2.$$

The vector $\mathbf{h}_i$ contains $N_R \times 1$ channel coefficients from the ith transmitting antenna. The anti-diagonal elements of matrix $\mathbf{U}$ satisfy

$$\mathbf{H}_1^H\mathbf{H}_2\mathbf{H}_2^H\mathbf{H}_1 = (\epsilon + \chi)\mathbf{I}$$

where $\epsilon = \left|\mathbf{h}_1^H\mathbf{h}_4 - \mathbf{h}_2^T\mathbf{h}_3^*\right|^2$ and $\chi = \left|\mathbf{h}_1^H\mathbf{h}_3 + \mathbf{h}_2^T\mathbf{h}_4^*\right|^2$.

It can be shown [135] that a Moore–Penrose channel inversion can be attained using linear processing exploiting orthogonal properties of the LSTBC and the estimated transmit vector $\hat{\mathbf{s}}$ can be written as

$$\hat{\mathbf{s}} = \frac{1}{k_1k_2 - (\epsilon + \chi)} \overbrace{\begin{bmatrix} k_2\mathbf{I} & -\mathbf{H}_1^H\mathbf{H}_2 \\ -\mathbf{H}_2^H\mathbf{H}_1 & k_1\mathbf{I} \end{bmatrix}^{-1}}^{\mathbf{F}_2} \overbrace{\begin{bmatrix} \mathbf{H}_1^H \\ \mathbf{H}_2^H \end{bmatrix}}^{\mathbf{F}_1} \mathbf{y} \tag{3.31}$$

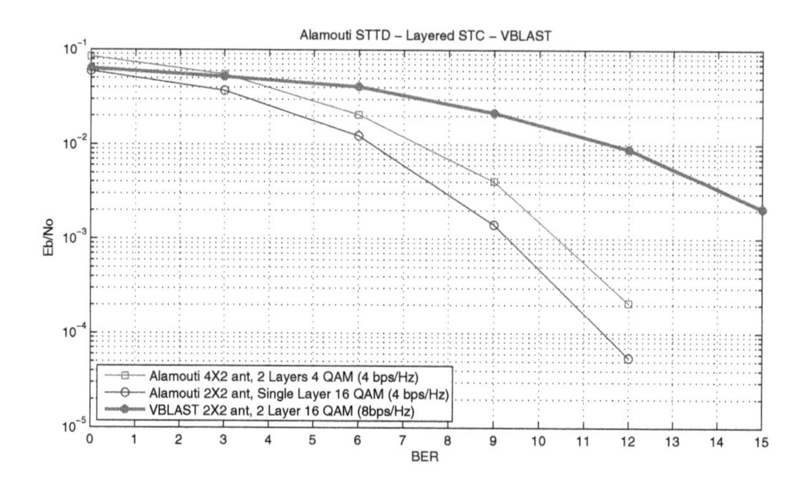

Figure 3.2 STTD and Layered STBC comparison

First, the receiver vertically processes two spatially multiplexed streams with matrix $\mathbf{F}_1$. Then, with matrix $\mathbf{F}_2$, it cancels cross-products (interferences). According to (3.31), SNR undergoes a reduction that is proportional to $(\epsilon + \chi)$. The equivalent SNR values for the estimation of the received symbols are

$$\mathrm{SNR}_{12} = \frac{k_1 - (\epsilon + \chi)}{k_2} \quad \text{and} \quad \mathrm{SNR}_{34} = \frac{k_2 - (\epsilon + \chi)}{k_1} \tag{3.32}$$

LSTBC offers the ability to transmit simultaneously to two different users at the cost of a BER degradation due to the spatial multiplexing of Alamouti blocks. This can be beneficial when multiple connections are active in a cell and strict QoS requirements are to be met, in particular regarding delay constrains, and a multi-user scheduling on the same time-frequency resource is desirable. Figure 3.2 reports the comparison for the BER curve over the SNR in dB for STTD and Layered STBC schemes.

3.4 MIMO Support in WiMAX

IEEE 802.16 standards [47, 48] support multiple antenna systems. In particular, 802.16e standard includes mobile support and adopts all the three solutions above: Alamouti, LSTBC and SM. Mainly due to their cost, antenna elements are principally deployed at the BS and receivers generally benefit of one or two antennas. The 802.16 multi-antenna support falls inside two

Table 3.1 IEEE802.16d/e MIMO support

	SC	SCa	OFDM	OFDMA
802.16d	no	STTD, AAS	STTD, AAS	2/4 BS antennas STTD and SM, AAS
802.16e	no	STTD, AAS	STTD, AAS	2/3/4 BS antennas, grouping, selection, STTD, LSTBC, SM, Collab. UL SM, AAS with AMC perm.

main classes: the Adaptive Antenna Systems (AAS) and the MIMO schemes applied to OFDM or OFDMA PHY layer.

IEEE 802.16-2004 (also known as IEEE 802.16d) standard [47] firstly introduced multi-antenna support. The "Amendment 2 and Corrigendum 1 to IEEE std. 802.16-2004", namely 802.16e version [48], added several details about AAS and MIMO [86].

The support of multiple antenna systems by IEEE802.16d/e is summarized in Table 3.1. The columns report the 802.16 PHY layer configurations: Single Carrier (SC), Single Carrier advanced (SCa), OFDM and OFDMA. The first two configuration operate in Single Carrier over a narrowband flat-fading channel. SC is design to work on the 10–60 GHz band, while the second configuration in the lower 2–11 GHz band. The difference between OFDM and OFDMA is mainly related to the allocation mechanism. In OFDM, a single OFDM symbol is to be directed to a single user, while in OFDMA the transmissions are based on frames. Each frame is divided in smaller resource units: time-frequency bins (or chunks), which can be assigned to different users. AAS is briefly described in the next section.

3.4.1 Adaptive Antenna Systems (AAS)

The Adaptive Antenna Systems (AAS) scheme works as a beamforming technique. In AAS, the signals from several antenna elements (not necessarily a linear array) are weighted (both in amplitude and in phase) and combined to maximize the Signal to Interference Ratio (SIR) at the receiver side. AAS is used only in the OFDMA PHY layer configuration. The smart antennas beams are not fixed and can place nulls in the radiation pattern to cancel interferences and mitigate fading, with the objective of increasing SIR and the SE of the system.

The IEEE 802.16d standard defines a signalling structure that enables the use of adaptive antenna system. A Point to MultiPoint (PMP) frame structure is defined for the transmission on downlink and uplink using directional

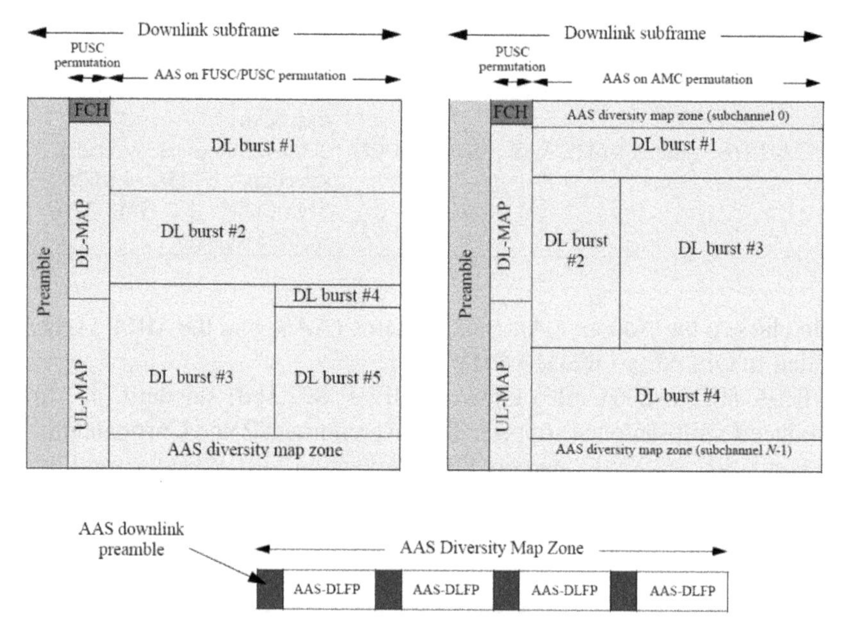

Figure 3.3 AAS frame structure in IEEE802.16, reproduced from [47]

beams, each one covering one or more MTs. The BS forms a beam based on channel quality reported by the MTs.

The part of the frame with adaptive antenna transmission is sent in a dedicated AAS zone, which spans over all the subchannels until the end of the frame or until the next permutation zone, as shown in Figure 3.3. AAS support for AMC permutation has been added in IEEE 802.16e version of the standard.

When a Discrete Fourier Transform (DFT) size greater than or equal to 512 is used, the BS can decide to allocate an AAS Diversity-Map Zone. The Diversity-Map Zone is placed accordingly to the permutation used. In Partially Used Subcarriers (PUSC), Fully Used SubCarriers (FUSC), and optional FUSC permutation, it is positioned on the two highest numbered subchannels of the DL frame, while in AMC it is positioned on the first and last subchannels of the AAS Zone [48].

These subchannels will be used to transmit the AAS-DownLink Frame Prefix (DLFP) whose purpose is to provide a robust transmission of the required BS parameters to enable MT initial ranging, as well as the beam selection via beam indexes and the resource assigned in the AAS zone.

In order to enter the network involving the DLFP, an AAS-MT follows a specific procedure. The Downlink Channel Descriptor (DCD) and Uplink Channel Descriptor (UCD) offer information useful for decoding and demodulation. A channel with broadcast Connection IDentifier (CID) is readable by every MT. The channels are allocated accordingly to the DL/UL maps transmitted after the frame header. An MT entering the network in AAS mode follows these steps:

- The AAS-MT synchronizes time and frequency by using the DL preamble.
- After synchronization, AAS-MT can receive necessary messages to identify: used modulation and coding (as the DCD and UCD pointed to by the allocation specified in the AAS-DLFP using the broadcast CID).
- Then, after the AAS-MT has decoded the DCD and UCD, the AAS-MT should perform ranging.
- The AAS-MT can receive a ranging response message through a DL-MAP allocation pointed to by an AAS-DLFP with the broadcast CID.
- Then AAS-MT can receive initial downlink allocations through a DL-MAP allocation pointed to by the AAS-DLFP with broadcast or specific CID.
- Other allocations can be managed by private DL-MAP and UL-MAP allocations.

3.4.2 Multiple Input Multiple Output System (MIMO)

MIMO support has been introduced in WiMAX since the IEEE 802.16d version has been released. In IEEE 802.16e, several features have been added for OFDM and OFDMA PHY layers. These systems can be used to achieve diversity gain in a fading environment or to increase capacity.

When transmit diversity is desired, multiple copies of the same data stream are transmitted over independent spatial channels which are created by employing multiple antennas. Transmission is more robust to wireless channel fluctuations, since it is unlikely that all the channels will fade simultaneously.

When higher capacity is needed, spatial multiplexing is utilized to transmit various streams of data simultaneously over different antennas in the same time slot, over the same frequencies. If statistical decorrelation among antenna elements is available, multiple transmit and receive antennas can create independent parallel channels and the transmitted symbols can be correctly reconstructed at the receiver. By using antennas well separated in space – by more than $\lambda/2$ and/or with different polarization – the decorrelation condition can be satisfied.

IEEE 802.16e standard defines three MIMO operation modes: Alamouti Code, referred to as STTD, as in Section 3.2.1 and in [5], LSTBC, as described in Section 3.3.1, and VBLAST SM, see Section 3.3 and [26,30].

Adaptive selection among the previous MIMO schemes, jointly with adaptive modulation and coding techniques, can offer high flexibility at the PHY layer maximizing data throughput and coverage.

Alamouti STTD is standardized for SCa, OFDM, and OFDMA PHY Layers. The receiver is a linear combiner [5] where symbols can be reconstructed by using orthogonal properties of space-time coding matrix **A** as in (3.4).

For SCa (Single Carrier for NLoS operation in the frequency bands below 11 GHz), STTD is used at a burst level. A burst p, composed by QAM symbols, has length F as in Figure 3.4. The bursts are arranged on a time basis. The STTD requires the processing of a pair of time bursts. A Cyclic Prefix (CP) is pre-appended to absorb the effect of signal dispersion due to the multipath: the length of the CP must be longer than the root-mean-square of the channel delay spread.

The first antenna transmits two sequence of F symbols, while the second antenna conjugates the transmitted complex symbols and time-reverses the sequence of data within each burst. The index n is the running position inside the burst. Due to the DFT/Inverse Discrete Fourier Transform (IDFT) operations, the bursts need also a time reversing operation which is realized by the $(F-n)_{\text{mod}(F)}$ operation, as in Figure 3.4, where the transmission for a pair of bursts (p_0 and p_1) is reported. A portion of the bursts (U symbols) is copied to form the CP. The receiver for the STTD scheme in SCa PHY layer can be found in [47, section 8.2.1.4.3.1].

In OFDM and OFDMA PHY layers, STTD transmission operates differently. OFDM and OFDMA are differentiated by the minimum data unit that can be manipulated. In OFDM PHY layer, Alamouti operates on two subsequent OFDM symbols composed by all the data subcarriers available in the system, while, in OFDMA, STTD can operate on a single group of subcarrier in the time-frequency allocation grid (a resource chunk).

In both cases, the symbols are coded accordingly to the matrix in (3.4): rows represent the transmitting antennas and columns represent OFDM symbols or subcarriers for OFDM/OFDMA PHY layer respectively. Compared to the SCa PHY layer case, no time-reverse operation is needed, since STBC coding is done on the single subcarrier and not on a full burst level.

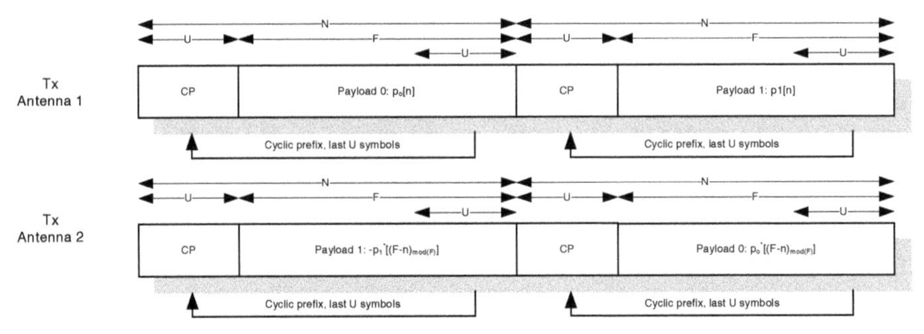

Figure 3.4 Alamouti transmission for SCa PHy layer profile

When the BS has three or four antennas, it is not possible to achieve a full-diversity approach, since it has been demonstrated that a full-rate, fully orthogonal Space-Time Code only exists for two antennas [42]. When a full rate transmission is desired, LSTBC schemes have to be used. Data rate is increased at the expense of one of the following: diversity gain, linearity or orthogonality. In IEEE 802.16e, the proposed scheme for four antennas is expressed by the coding matrix **B**, as in (3.28), thus orthogonality is lost.

As a third MIMO option, the system can switch on a VBLAST transmission [30]. Independent data streams are spatially multiplexed, i.e., they are transmitted from different antennas in the same OFDM symbol time. VBLAST follows the coding matrix **C** as in (3.27).

While Alamouti and its hybrid version (two layered Alamouti blocks) do not exploit all the freedom degrees of the MIMO channel, VBLAST can extract the complete capacity gain.

VBLAST receiver decodes the streams in successive steps as described in Section 3.3. Receiver algorithm starts from the detection of the signal with the highest SIR. The detected symbol is then cancelled from each one of the remaining received signals, and so on. It is evident the analogy with the multi-user interference cancellation. The drawback of every SIC method is that an error in the first stages of processing can propagate destructively to the successive steps of detection; for this reason, the maximum SNR signal is decoded first in order to lower the probability of error propagation. Refer to [130] for a comprehensive VBLAST performance evaluation.

In uplink transmission, an additional scheme is available: the *collaborative SM* scheme. In a multiuser collaborative SM context, two users may share the same channel in the UL. In this case, an MT can request an allocation in the uplink to be used in coordination with a second MT in order to perform a Collaborative Spatial Multiplex. MT participating at the Collaborative SM can be equipped with a single antenna. The BS will separate the streams transmitted by each MT.

3.4.3 Adaptive Modulation and Coding

In wireless systems, the signal transmitted to and by a station can be modified to counter-react to the signal quality variations through a process commonly referred to as *link adaptation*. This allows to improve system capacity, peak data rate, and coverage reliability. Traditionally, wireless systems use fast *power control* as the preferred method for link adaptation. In a system with power control, the power of the transmitted signal is adjusted in order to meet a target carrier-to-interference-plus-noise ratio at the receiver. Typically, the transmit power is low when a user is close to the BS and it increases when the user moves away from the BS.

Adaptive Modulation and Coding (AMC) offers an alternative link adaptation method. In a system with AMC, the power of the transmitted signal is held constant, but the modulation and coding formats are changed to match the current radio link quality. AMC provides the flexibility to match the modulation-coding scheme to the average channel conditions of each station. Users close to the BS are typically assigned higher-order modulations and high code rates. The modulation-order and/or the code rate are decreased as the distance from the BS increases.

In particular, in an OFDM/OFDMA wireless system, the inherent multicarrier nature of OFDM modulation permits the use of AMC according to the behavior of each narrow-band channel (subcarrier). Different subcarriers can be allocated to dishomogenous users to provide a flexible multiuser access scheme, that exploits multiuser diversity.

Adaptive Modulation and Coding mechanisms supported by IEEE802.16 are briefly explained; adaptive MIMO and AMC are jointly used in the dynamic MIMO *link adaptation* in Section 3.5.

Usually, the key metric for link adaptation is the BER and is related to the SE of the scheme used. In this sense, MIMO schemes will be adaptively changed combined with different modulation orders to vary the SE of the transmission based on the channel quality.

AMC is twofold supported by IEEE 802.16 standard [47]. First, a large selection of modulation and channel coding is available at the BS and at the MT. All the possible combinations between the modulations (QPSK,

16-QAM, and 64-QAM) and the coding rates (1/2 and 3/4) are allowed. In addition, the 64-QAM modulation can be also combined with a coding rate 2/3.

When AMC is used, a special permutation scheme is defined, where subchannels are composed by groups of contiguous subcarriers. In AMC permutation, the smallest allocation unit is called *bin* and consists of 9 contiguous subcarriers (eight data subcarriers and one pilot subcarrier) on a single OFDM symbol. Let N be the number of contiguous bins and M be the number of contiguous symbols. A slot in AMC is defined as a collection of bins of type $(N \times M)$, with $(N \times M) = 6$. Thus, the allowed combinations are: (6 bins, 1 symbol), (3 bins, 2 symbols), (2 bins, 3 symbols), and (1 bin, 6 symbols).

In OFDM/OFDMA systems, in addition to power control and AMC, *Dynamic Subcarrier Assignment* (DSA) can be used. Due to the contiguous subcarrier allocation, each user can experience highly-variable channel conditions and may benefit from multiuser diversity by choosing the subchannel with the best frequency response. High spectrum efficiency can be achieved with adaptive subchannel allocation strategies.

In conclusion, there is a wide degree of flexibility for radio resource management in the context of OFDM/OFDMA wireless systems. Since channel frequency responses are different at different frequencies for different users, adaptive power allocation, AMC, and DSA can significantly improve the performance of such systems. Link adaptation algorithms can be designed to maximize the overall network throughput or to achieve target error performance.

Dynamic link adaptation in the DL is obtained through MT feedback, by providing the transmitting BS with CSI estimates, as illustrated in Figure 3.5. In the uplink the MT can change its transmission parameters based on its own channel estimates, if a good channel reciprocity[7] can be assumed.

The accuracy of the channel estimates and the latency of the feedback affect the AMC algorithms performance. AMC strategies have to define how quickly the transmitter must change its parameters. Since the constellation size is adapted to an estimate of the channel fade level, several symbols may be required to obtain a good estimate. These practical considerations are relevant in a mobile context with fast-varying channels and require accurate solutions such as prediction models for the channels.

---

[7] A radio channel between a BS and a MT is said to be fully reciprocal if the channel sensed by the MS is equal to the channel sensed by the BS.

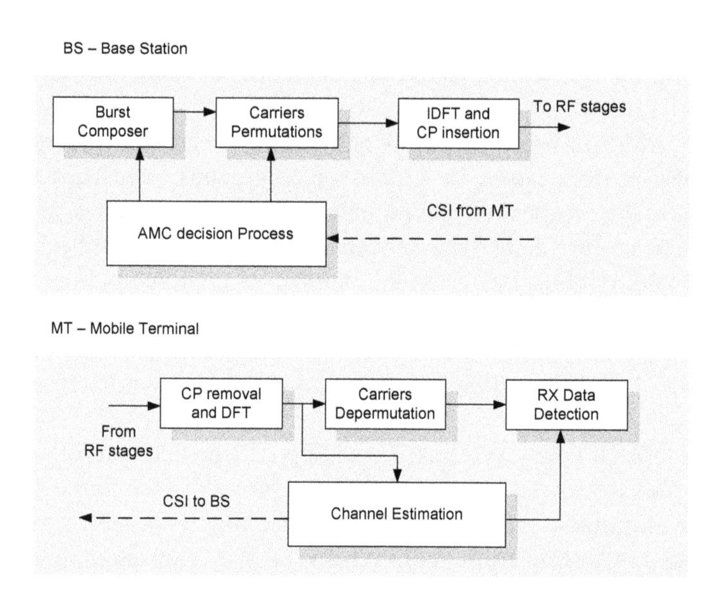

Figure 3.5 IEEE802.16 AMC system diagram

The MU MIMO-ARQ system proposed in Chapter 5 will address the problem of fast feedback of the CSI that is challenging for power-control and AMC schemes. A cross-layer ARQ state machine will be designed that do not need any CSI feedback. An opportunistic multiuser scheduling will be shown to attain performance improvements comparable with an informed user selection method based on the selection of the user with the best post-processing SNR.

3.5 Link Adaptation with Dynamic MIMO

An adaptive selection of MIMO schemes is proposed considering the performance of STTD, LSTBC and SM. A selection of different transmission schemes based on the CSI increases the Spectral Efficiency (SE) while maintaining the BER under a specified threshold. Link adaptation is achieved via AMC and switching of various MIMO schemes (DIV/SM tradeoff).

The tradeoff has been analyzed in [113] from an information theoretic point of view with an analysis for high SNRs. It has been demonstrated that spatial and diversity gains can be achieved simultaneously. The work quantifies the amount of diversity and multiplexing gains with ideal coding and the tradeoff between them depending on the number of antennas of the MIMO system.

With some hypothesis, this tradeoff can be simply expressed. Consider a $N_t \times N_R$ MIMO system transmitting over an identically independent distributed (iid) spatially white Rayleigh flat-fading channel; consider the channel constant over the coding block length l:

$$l \geq N_T + N_R - 1$$

and a coding scheme with a multiplexing gain equal to r. The optimal diversity gain achievable $d^*(r)$ can be expressed as

$$d^*(r) = (N_T - r)(N_R - r) \tag{3.33}$$

The antenna resource, in the case of ideal coding and constant channel over the block code, can be independently exploited for multiplexing or diversity gains. Out of the total $N_T \times N_R$ antenna, r are used for spatial multiplexing while the remaining $(N_T - r)$ at the transmitter and $(N_R - r)$ at the receiver are used for diversity. The multiplexing-diversity tradeoff represents the tradeoff between the data rate and the error probability of a MIMO system.

An optimal tradeoff curve can be plotted varying the multiplexing gain from zero towards its maximum: $0 \mapsto \min(N_T, N_R)$. The correspondent diversity gain varies from $0 \mapsto N_T \times N_R$, as illustrated in Figure 3.6.

3.5.1 MIMO Switching Based on Euclidean Distance

The previous analysis with an ideal coding of infinite length, and for SNR $\rightarrow \infty$ gives an useful tool to compare existing MIMO coding schemes but has no direct implication in the design of adaptive MIMO systems.

Several designs for MIMO adaptivity have been proposed. In [74, 76], switching algorithms between diversity and spatial schemes were proposed based on theoretical BER; in [124] the effect of partial channel estimation for dynamic MIMO systems was considered.

In [41], a practical system for DIV/SM dynamic switching among equal SE MIMO schemes was proposed which considered a narrowband system with a Rayleigh channel. The adaptive criterion is based on the *Demmel condition number* of the channel matrix. When equal SE schemes are used, it is meaningful to assess a criterion based on the Euclidean Distance (ED) of the codebook (or constellation).

The minimum codebook ED metric can be expressed for MIMO multiplexing and diversity codes; based on instantaneous channel condition, a

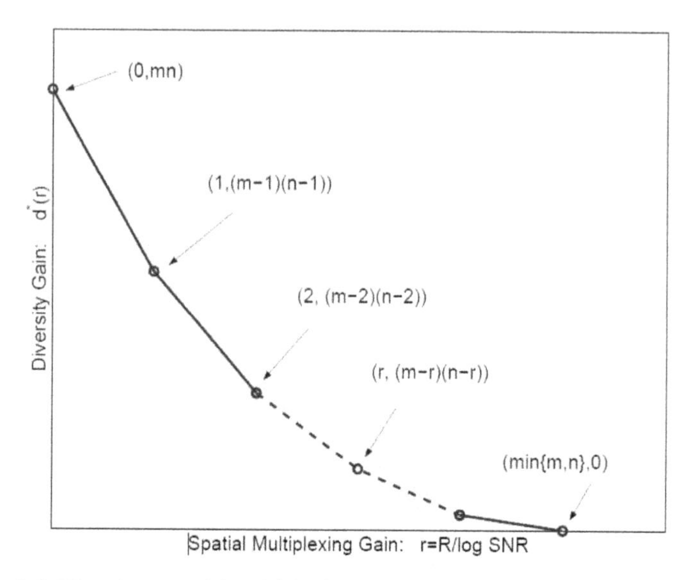

Figure 3.6 Diversity – spatial multiplexing tradeoff, $m = N_T$ and $n = N_R$, from [113]

decision can be performed in order to adapt to the requirements in terms of symbol and error rates. Let us define the following codebook distances:

- $d^2_{\min,SM}$: the squared ED for SM at the source
- $d^2_{\min,MD}$: the squared ED for DIV at the source
- $d^2_{\min,sm}$: the squared ED for SM at the receiver
- $d^2_{\min,md}$: the squared ED for DIV at the receiver.

For the SM schemes, the codebook ED at the receiver is related to the channel eigenvalue distribution, where each eigenvector corresponds to the energy of a spatial path in the channel. It can be demonstrated that for SM, the ED at the receiver is bounded as

$$\lambda^2_{\min}(\mathbf{H})\frac{d^2_{\min,sm}}{N_t} \leq d^2_{\min,SM}(\mathbf{H}) \leq \lambda^2_{\max}(\mathbf{H})\frac{d^2_{\min,sm}}{N_t} \tag{3.34}$$

where $\lambda_{\min}$ and $\lambda_{\max}$ are the minimum and maximum eigenvalues of the MIMO channel matrix $\mathbf{H}$ and N_T is the number of transmitting antennas.

For the diversity schemes, the distance at receiver is directly proportional to the number of transmit antennas and inversely proportional to Frobenius norm of the matrix. The following relation holds:

$$d^2_{\min,MD} \leq \frac{1}{N_t}\|\mathbf{H}\|^2_F d^2_{\min,md} \tag{3.35}$$

where $\|\mathbf{H}\|^2_F$ is the squared Frobenius norm of the MIMO channel matrix.[8]

The Frobenius norm (or the Hilbert-Schmidt norm) can be defined in various ways:

$$\begin{aligned}
\|\mathbf{H}\|_F &= \sqrt{\sum_{i=1}^{N_t}\sum_{j=1}^{N_r}|h_{ij}|^2} \\
\|\mathbf{H}\|_F &= \sqrt{\mathrm{Tr}(\mathbf{H}^\dagger \mathbf{H})} \\
\|\mathbf{H}\|_F &= \sqrt{\sum_{i=1}^{\min(N_T,N_R)} \lambda_i^2}
\end{aligned} \tag{3.36}$$

From (3.34) and (3.35), it is clear that a criterion can be derived for MIMO scheme selection. Fixing the same transmit codebook ED at the receiver and referring to the lower bound in (3.34), a conservative criterion for choosing SM can be derived. SM is chosen when the following relation holds:

$$\lambda^2_{\min}(\mathbf{H})\frac{d^2_{\min,sm}}{N_T} \geq \frac{1}{N_T}\|\mathbf{H}\|^2_F d^2_{\min,md} \tag{3.37}$$

The previous can be rewritten using the definition of *Demmel Condition Number* of the matrix **H**. The Demmel Condition Number is defined as[9]

$$K_D = \|\mathbf{H}\|_F / \lambda_{\min}(\mathbf{H}) \tag{3.38}$$

Thus, Eq. (3.37) can be expressed as

$$\|\mathbf{H}\|^2_F d^2_{\min,md} \leq \lambda^2_{\min}(\mathbf{H}) d^2_{\min,sm} \tag{3.39}$$

and using the definition of the Demmel condition number:

$$K_D \leq \frac{d_{\min,sm}}{d_{\min,md}} \tag{3.40}$$

[8] For Alamouti, Equation (3.35) holds with equality.

[9] The "standard" condition number can be calculated as $\|\mathbf{H}\|\|\mathbf{H}^{-1}\|$ for any order of the matrix norms; in the case of Demmel condition number, it is calculated as $K_D = \|\mathbf{H}\|_F\|\mathbf{H}^{-1}\|_2$.

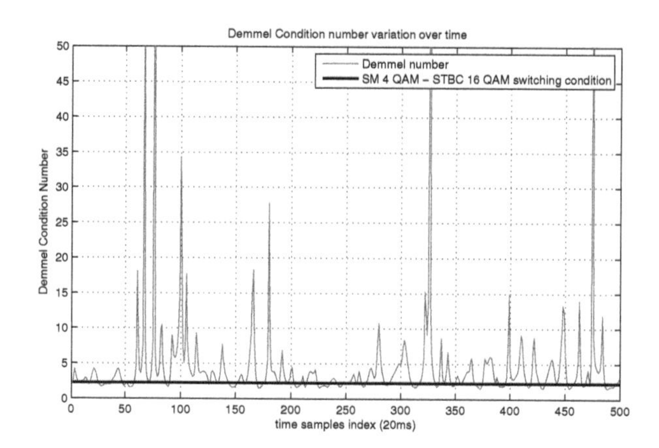

Figure 3.7 Demmel condition number simulated for WINNER II channel and SM-STBC switching condition as in [41]

The DIV/SM criterion can be summarized as follows: *the SM scheme is chosen if the ratio on the right-hand side of (3.40) is greater than the MIMO channel Demmel condition number* K_D.

If small ratios of $d_{\min,md}/d_{\min,sm}$ are accepted by design, the criterion allows to exploit SM with lower scattering channels with the penalty of higher BERs.

In Figure 3.7, the Demmel condition number measured over time is reported along with the SM-STBC switching condition calculated for 4-QAM SM (VBLAST) and 16-QAM STBC (Alamouti). The channel simulation is based on WINNER II channel model. The time sampling period is 20 ms and the link is simulated in Urban Macro Cell (scenario C2) for a distance between BS and MS of 1200 m and a velocity of 120 Km/h. The channel simulator is available at [60]. The C2 channel model is described in [61]. It is evident that the switching condition can change several times per second: this requires a very fast adaptive mechanisms which is able to switch from diversity to spatial multiplexing schemes with very low delay and hysteresis.

The switching condition C_{SW} is calculated considering the minimum distance among symbol in M-QAM modulation [90], which is

$$d^2_{\min,M} = \frac{12}{2^M - 1} \tag{3.41}$$

Evaluating for $M = 2$ (4-QAM) and $M = 4$ (16-QAM), the following is obtained:

$$C_{sw} \leq \sqrt{\frac{d^2_{\min,2}}{d^2_{\min,4}}} = \sqrt{\frac{4}{4/5}} = \sqrt{5} \quad (3.42)$$

If $C_{SW} \geq K_D$, the SM can be used, or, equivalently, SM can be used if the condition number is lower than $\sqrt{5}$.

3.5.2 Adaptive MIMO with Upper-Bounded BER

A selection of different multi-antenna systems is analyzed in this section with the aim of keeping an upper-bounded BER over a wide range of SNRs while maximizing the SE, as reported in [74]. An adaptive selection is performed based on BER of the multi-antenna schemes proposed in the IEEE802.16e standard, also considering channel estimation issues. Sensitivity to channel estimation is analyzed using a channel estimator operating with linear interpolation only in the frequency domain.

The channel estimator is based on pilots transmitted according to the adjacent data-pilot permutation. Channel estimator feeds the multi-antenna receivers and provides a CSI measure to select the correct AMC profile and MIMO scheme.

The adjacent data-pilot permutation, or AMC permutation, is implemented as defined in [47,48]. The choice of this permutation permits to modulate and code the chunks independently on the basis of CSI measured by the MT.

In Figure 3.8 a 4×2 MIMO system is shown. The BS receives from the MT a control signal to operate the best selection of MIMO transmission scheme. The return link is supposed to be delay-less and error-free. The MIMO encoded streams are allocated on the OFDM symbols which are composed of encoded data, pilots, left and right guard subbands and a central DC tone. Data and pilots subcarriers are ordered according to AMC permutation. More details about IEEE802.16e AMC permutation are reported in Section 3.5.3. A CP is added to avoid Inter-Symbol Interference (ISI). The length of CP is $L_{CP} > \Theta - 1$ where Θ is the maximum length of the channel impulse response.

At MT, after CP removal and FFT, pilots are extracted to obtain channel estimation. Channel coefficients and data subcarriers are fed into MIMO receivers. The correct receiver is selected on the basis of the indication provided by the BS in the downlink map. Channel estimator operates a linear interpolation over the frequency domain, as detailed in Section 3.5.3.

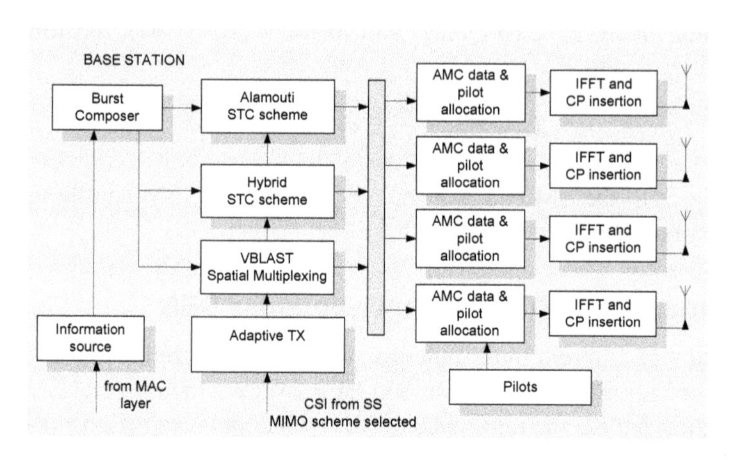

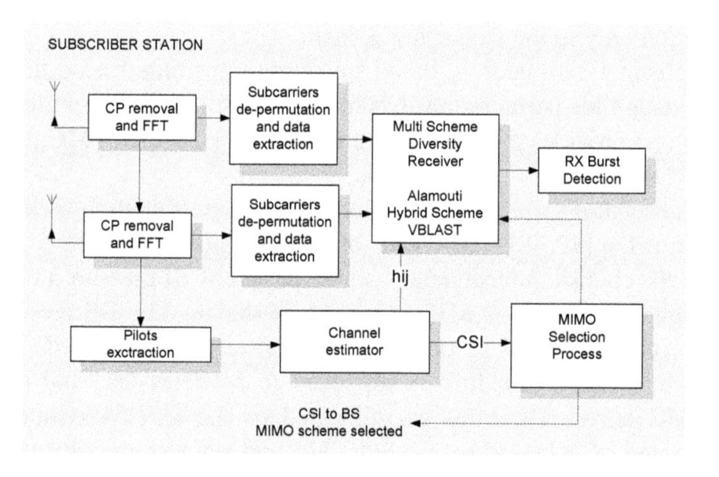

Figure 3.8 Functional blocks of the considered BS and MT

Received signal at time n for each subcarrier can be expressed as

$$\mathbf{y}_n = \mathbf{H}_n \mathbf{s}_n + \mathbf{n}_n \tag{3.43}$$

where $\mathbf{H}_n \in \mathbb{M}_{N_T \times N_R}$ is the channel matrix, $\mathbf{s} = [s1, s2, s3, s4]^T$ is transmitted symbol vector and vector $\mathbf{n}$ is additive Gaussian noise with zero mean and variance σ^2.

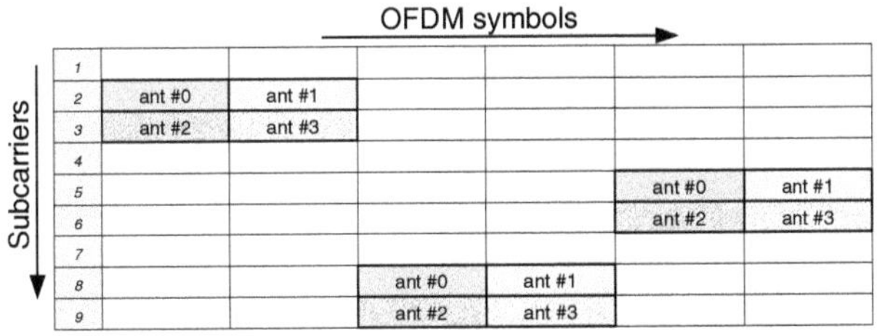

OFDM symbols →

Subcarriers ↓

1						
2	ant #0	ant #1				
3	ant #2	ant #3				
4						
5					ant #0	ant #1
6					ant #2	ant #3
7						
8			ant #0	ant #1		
9			ant #2	ant #3		

Figure 3.9 AMC permutation pilots position in 4×4 multi-antenna system

3.5.3 Pilot Structure and Channel Estimation

IEEE802.16 PHY layer considers different data and pilots permutations. In PUSC and FUSC modes data positions are scrambled on the whole OFDM symbol, thus channel state is averaged in the frequency domain and upper layers cannot exploit heterogeneous link qualities with Dynamic Subcarriers Assignment. AMC permutation, instead, groups data subcarriers in adjacent blocks with heterogeneous channel qualities.

AMC permutation allows the MAC layer to chose independent profiles for each burst in order to match QoS and upper-layers requests.

According to Section 3.4, in this work the $6 \cdot 1$ configuration is used which is allocated only in the frequency domain.

Pilots are positioned in linear ascending fashion on a regular grid, as in Figure 3.9, where three bins are shown. The processing of regular spaced grids can be performed as indicated in [77].

The adopted channel estimator operates a linear interpolation over the frequency dimension to extract channel coefficients for each data subcarriers, no time-domain correlation is exploited and a new channel estimation is performed for each OFDM symbol. Channel estimation has a two-fold role: first it feeds the equalizer with frequency domain channel coefficients for decoding process, then provides a CSI measure helpful to select the correct AMC profile and MIMO scheme. Pilots from multiple antennas are multiplexed; when an antenna is transmitting its pilot tone, the other three antennas are switched off thus allowing a simple reconstruction of MIMO channel coefficients, as visible in Figure 3.9. Feedback to the transmitter is here considered delay-less and with perfect quantization.

Table 3.2 Upper-bounded BER MIMO switching parameters

Parameter	Value
OFDM useful time	89.6 μs
CP time	11.2 μs
CP Ratio	1/8
FFT size	1024
Subcarrier frequency spacing	11.16071429 kHz
Guard subcarriers	159
Used subcarriers	865
Data subcarriers	768
Pilot subcarriers	96
Frame duration	2 μs
Per frame OFDM symbols	19 (1 preamble)

The position of the pilots inside the symbol, in the case of $N_T = 4$, is ruled by the following equations:

$$\begin{aligned}
\text{Pilot Location}|_{\text{ant0}} &= 9k + 3\, m_{\text{mod3}} + 1 && m \text{ even} \\
\text{Pilot Location}|_{\text{ant1}} &= 9k + 3\, (m-1)_{\text{mod3}} + 1 && m \text{ even} \\
\text{Pilot Location}|_{\text{ant2}} &= 9k + 3\, m_{\text{mod3}} + 2 && m \text{ odd} \\
\text{Pilot Location}|_{\text{ant3}} &= 9k + 3\, (m-1)_{\text{mod3}} + 2 && m \text{ odd}
\end{aligned}$$

where m is the time index and k is the subcarrier index.

3.5.4 Performance Analysis

Computer simulation results for the different STBC, LSTBC and SM schemes considered in the IEEE802.16e standard are presented. A spatially white MIMO Rayleigh fading channel with iid coefficients is considered. The simulated system selects a MIMO transmitting scheme accordingly to the estimated SNR. The communication is in Point to Point (PtP) from the BS to the MT.

Simulation parameters have been chosen accordingly to IEEE 802.16e standard. Table 3.2 reports the main parameters used in the simulation.

The channel is modeled according to ITU Pedestrian channel A, as in Table 3.3. The transmission modes available (as described in the previous sections of this chapter) are summarized in Table 3.4, and consist in all the combination of the three scheme with the three order of the QAM modulations. The diversity d and spatial r gains of the schemes are also reported.

Table 3.3 PDP for ITU pedestrian Type A multipath channel

Delay		**Gain**	
0	ns	0	dB
110	ns	-9.7	dB
190	ns	−19.2	dB
410	ns	-22.8	dB

Table 3.4 Upper-bounded BER MIMO switching: transmission modes

	Parameter Value				
MIMO modes	**Mod**	***d***	***r***	**Ant.**	**SE bit/s/Hz**
A	4 QAM	4	1	2×2	2
A	16 QAM	4	1	2×2	4
A	64 QAM	4	1	2×2	6
B	4 QAM	2	2	4×2	4
B	16 QAM	2	2	4×2	8
B	64 QAM	2	2	4×2	12
C	4 QAM	1	2	2×2	4
C	16 QAM	1	2	2×2	8
C	64 QAM	1	2	2×2	12

In Figure 3.10, the Alamouti scheme with $N_R = 2$ and the hybrid LSTBC scheme with $N_T = 4$ and $N_R = 2$ are compared. Both schemes obtain the same bandwidth efficiency of 4 bit/s/Hz. In the LSTBC, the loss of orthogonality is paid with a reduction of E_b/N_0 values, at fixed BER, due to interference from spatially multiplexed Alamouti blocks. The E_b/N_0 degradation is around 2 dB in the range from 2 to 12 dB. Despite of this performance loss, this scheme allows to allocate multiple user on the same time-frequency resource.[10]

A first adaptive system has been simulated with the constrain of a minimum spatial gain of 2 that realizes a selection of the adopted MIMO architectures allowing an increase in bandwidth efficiency. The scheme which provides a diversity gain of 2 are LSTBC and SM. The selection is performed in order to maintain the uncoded BER within the operational range of $8 \cdot 10^{-4} < \text{BER} < 10^{-2}$, as in Figure 3.11. The adaptive scheme aims at the highest spectral efficiency and will not use LSTBC. In the high SNR range, a SE of 8 bit/s/Hz is achieved; at 8.3 dB the system doubles efficiency, from 4 to 8 bit/s/Hz.

[10] Multi-user spatial multiplexing will be the key of the MIMO-ARQ protocol proposed in Chapter 5.

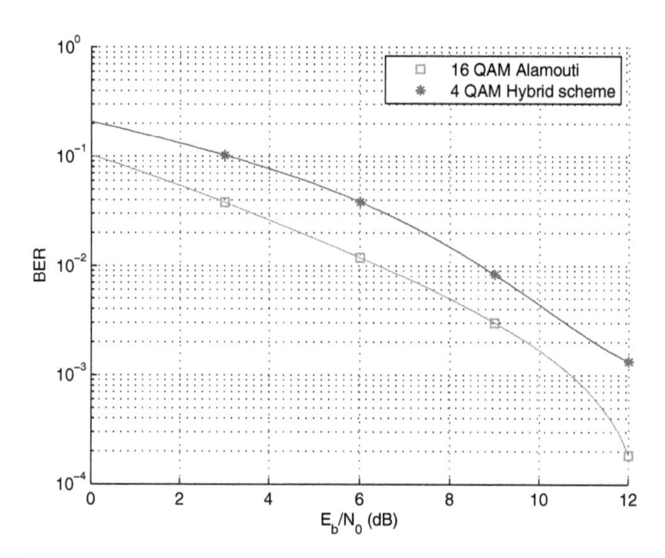

Figure 3.10 Alamouti and hybrid STC schemes comparison at low E_b/N_0

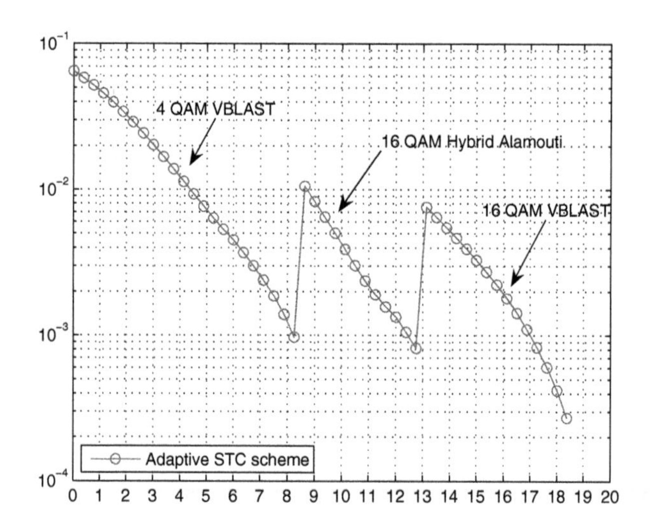

Figure 3.11 BER over SNR (dB) of a constrained adaptive system with multiplexing gain of 2, ITU vehicular channel A

If no diversity constrains are applied and the unique adaptation criterion is the maximization of the system SE with a constrained BER, LSTBC will not be chosen. This situation is illustrated in the following results which report uncoded bit error rates for Alamouti and SM MIMO schemes. Modulations used are 4–16–64-QAM.

Figures 3.12–3.13 show uncoded BER for Alamouti 2×2 and VBLAST 2×2 schemes with respect to the SNR. The SE is reported. The target BER is set to 0.01.

As shown in Figure 3.12, it is evident that in the case of flat fading channel, diversity schemes are chosen for any order of QAM modulation before resorting to spatially multiplexed transmissions and VBLAST 4QAM is never used. VBLAST SM has very poor performance at low SNR since the scheme needs a multipath environment to ensure multiple independent paths from the transmitter to the receiver. If a non-selective channel is encountered, the BLAST receiver tends to propagate errors in the decoding stages, leading to high bit error rates.

In the case of ITU multipath channel, Figure 3.13, BER performance of 16-QAM Alamouti and 4-QAM VBLAST schemes (same SE) are equal. In this specific case, that ranges from 6 to 10 dB, the selection may depend on other requirements, such as multi-user allocation on the same radio resource or independent ARQ acknowledgements, as later illustrated in Chapter 5.

A general remark is to use SM for higher data throughput at high SNR regions as long as the BER is below a desired threshold. When the channel quality is detected to be poorer, data transmission should be switched to exploit spatial diversity so that the reliability can be increased at low SNR regions: operative SNR thresholds need to be determined to switch among AMC/MIMO schemes.

However, when high throughput is requested and the channel quality is sufficient to support it with a specified BER, data transmission can switch to SM schemes with high order modulations increasing the overall data rate.

Throughput graphs have been also produced. In Figure 3.14, the throughput for an upper-bounded BER of 0.01 is shown in the case of ideal channel estimation. As expected, it can be seen that LSTBC has always a lower throughput respect to the equal efficiency (doubled modulation order) of the Alamouti scheme due to the imperfect orthogonality of the code. Between 16-QAM Alamouti and 4-QAM LSTBC, there is around a 2 dB loss and more than 1 dB between 64-QAM Alamouti and 16-QAM LSTBC. The VBLAST schemes start having an acceptable performance at higher SNRs reaching the target BER for 64-QAM only at around 23 dB.

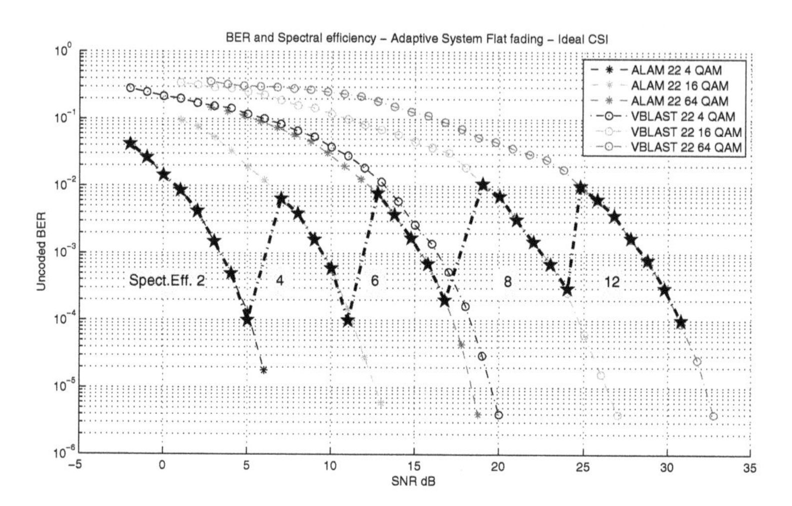

Figure 3.12 Adaptive MIMO link adaptation for upper-bounded BER, flat fading Rayleigh channel

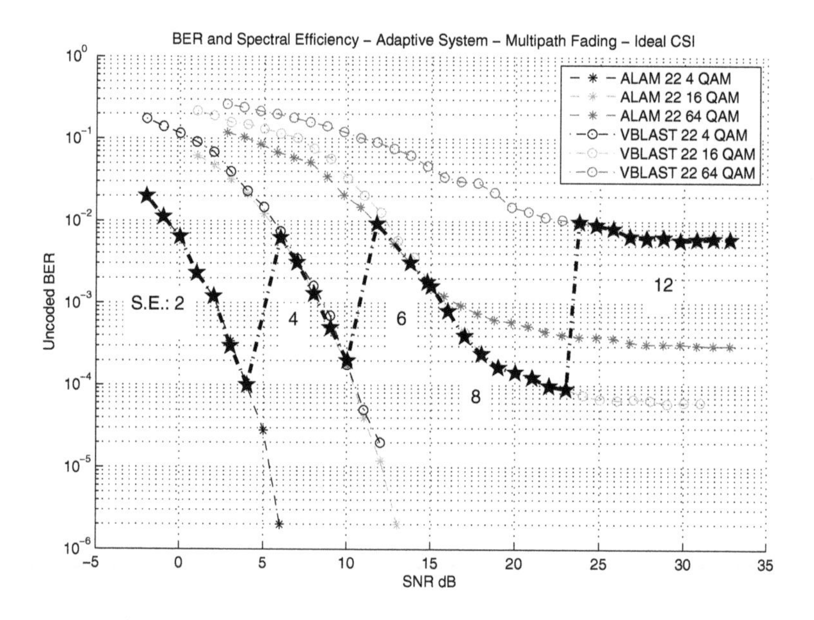

Figure 3.13 Adaptive MIMO link adaptation for upper-bounded BER, ITU vehicular channel A

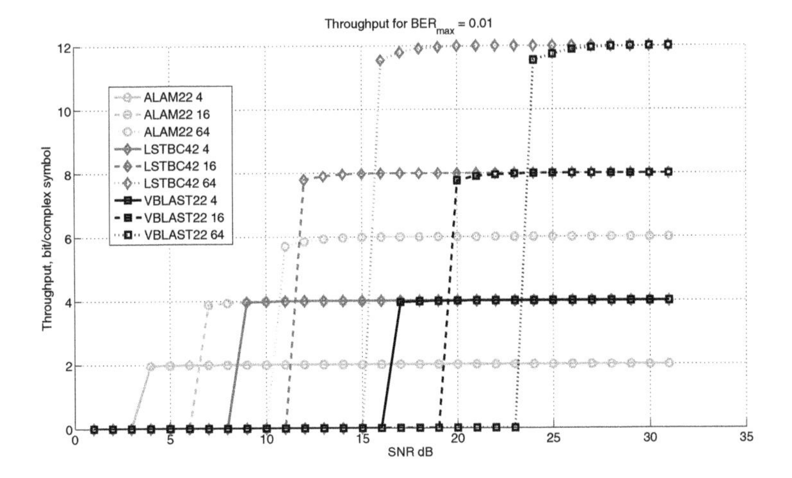

Figure 3.14 Throughput for ideal channel estimation for a $\mathrm{BER}_{\max} = 0.01$.

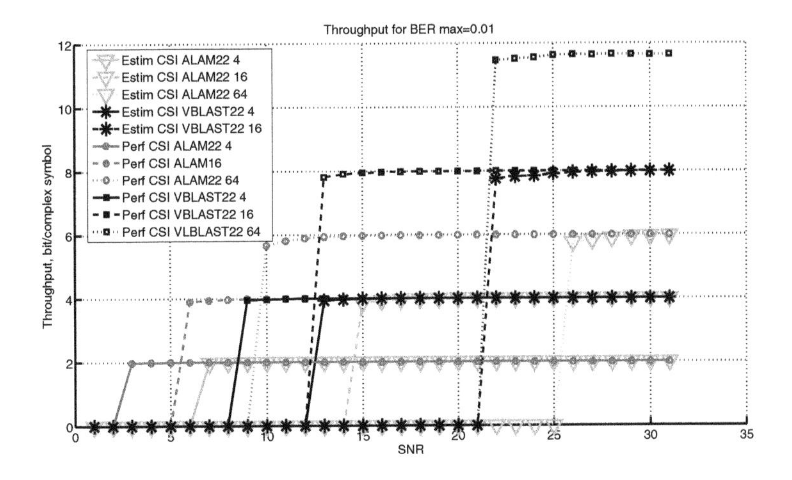

Figure 3.15 Influence of channel estimation over throughput for Alamouti and spatial multiplexing schemes

In Figure 3.15, the throughput of Alamouti and VBLAST schemes is compared in the case of a channel estimators operating as detailed in Section 3.5.3. The reference channel is ITU multipath channel (specified in Table 3.3 with a MT velocity set at 5 km/h. It can be noted that the channel estimation needs to be accurate in order to have good throughput performance. The simple interpolator in the frequency domain seems to be too far not enough robust to support the MIMO schemes analyzed; the 4-QAM Alamouti scheme has around 4 dB loss. At 16-QAM the dB loss is around 7 dB.

Based on their characteristics, the MIMO schemes adopted in IEEE 802.16e standard are selected in order to maximize the system capacity while having a BER under a fixed threshold. If no constrains are posed on the minimum multiplexing gain, the system maximizes the throughput. The system is proven to obtain a constrained BER over a wide SNR range. The sensitivity of MIMO schemes regarding channel estimation has also been reported for a linear estimator in the frequency domain.

3.6 Dynamic Space-Time-Frequency Block Codes

OFDMA and MIMO are the key technologies for the Radio Access Technology (RAT) in 4G networks. They are suitable to be applied together; the keyword MIMO-OFDM has been around in these last years as if their respective advantages were naturally born to inter-work together. A panorama over the challenges and technical solutions proposed in the field of MIMO-OFDM can be found in [102].

In OFDMA, the division of the wide-band channel in several subcarriers permits to allocate data in time and/or frequency domains with the opportunity for optimalization of STBCs. The three dimensions of time, frequency and antennas can be exploited together for allocation of Space-Time-Frequency Block Codes (STFBCs).

In the following part, the STFBCs proposed in the literature are briefly reviewed and then the proposed approach is described: a dynamic allocation of STBC and SFBC is studied. The method is based on a single-bit feedback and allows to counterfight the performance degradation due to the MT mobility and channel frequency selectivity. The dynamic allocation criterion is assessed for WSSUS channels.

3.6.1 Space-Time Frequency Block Codes (STFBCs)

In order to combine the advantages of the MIMO systems and OFDM, STFBC coded MIMO-OFDM systems have been designed in which coding is done across space, time and frequency to achieve maximum possible diversity gain. It has been proved that MIMO transmissions over a frequency-selective fading channels can provide a diversity gain which is, at maximum, the product of the number of transmit antennas, the number of receive antennas and the channel length, which is related to the channel Root Mean Square (RMS) delay spread.

It has been shown, in [11], that space-frequency-only coded OFDM systems cannot achieve full diversity: coding over the three space, time, and frequency dimensions is needed. In [51] a full-diversity STFBC is proposed using the concept proposed in [56] of Co-Ordinated Interleaved Orthogonal Designs (COID).

Three dimension coding has very complex implementation and simpler Space-Time-Frequency (STF) solutions have been proposed in literature. In [71], constellation precoding and Space-Time (ST) are used. The OFDM subcarriers frequency decorrelation can be exploited in order to preserve (and enhance) the diversity when the channel spatial correlation is high (thus the channel matrix rank is low). This can be accomplished using subcarriers grouping as in [71,83] or multiple channel fading blocks as proposed in [131]. In [83] subcarriers with independent fading are chosen achieving the highest frequency separation via a specific subcarriers permutation. A permutation is needed since the subcarrier frequency separation ΔF is designed in order to have high correlation among adjacent subcarriers, for channel estimation purposes [114]. Designs in [56, 71] are no rate-one orthogonal design for more than two transmit antennas. A rate-one STFBC for four transmit antenna MIMO-OFDM systems is presented in [37].

The solution proposed here [81] relies on a single-bit feedback for choosing between STBC and SFBC. The dynamic allocation in OFDM can be exploited to optimize the reception of MIMO streams for mobile MTs. The effect of the frequency selectivity and time variation of the wireless channel over STBCs in an OFDM system for a mobile MT is studied.

In order to mitigate the influence of the channel variability, a dynamic allocation over the time-frequency OFDM grid is proposed. The criterion for dynamic allocation direction is derived for a WSSUS channel.

It is shown that medium to high velocity MTs (50–300 km/h) have comparable correlation coefficients along time and frequency directions and results show that mobile users can greatly improve BER performance with dynamically allocated STFBC symbols. The SEP gain is analytically derived and computer simulations are provided for an ITU Vehicular channel for a DL connection in a IEEE802.16e system. In the time dimension, with fast-moving MTs, the channel has high variability. The movement of the MT causes variations in the frequency domain due to varying multipath delay spread and in the time domain due to the Doppler effect. The time variability is worsen in the higher spectrum bands.[11]

A dynamic allocation of the STBC data symbols along the time or frequency direction with a single-bit feedback is proposed aiming to maximize the correlation between successive STBC symbols. It is shown that STBCs can maintain the diversity if dynamically allocated. The effect of the channel multipath fading and Doppler spectra over STBCs in an OFDM system is evaluated. The dynamic allocation of STBC proposed has low impact on the legacy OFDM systems already developed and can enhance upper-layer performance leveraging the throughput at the interface between PHY and MAC layer [80].

Based on WSSUS channel model [9], the correlation coefficients are derived among subcarriers – $\rho_f(\Delta f)$ – and between different signalling times – $\rho_t(\Delta t)$. The correlation coefficients are used to analytically calculate the SEP expression for the Alamouti-coded symbols for any order QAM modulations. The STBC symbols are allocated based on whether correlation coefficient is the highest, between ρ_t and ρ_f, tracking the variation for the channel Power Delay Profile (PDP) and Doppler variations. Antennas are considered uncorrelated both at the MT and at the BS.

In Section 3.6.2, the OFDM system model is explained. In Section 3.6.3, the channel model is presented: correlation coefficients along time and frequency domains are derived based on PDP and terminal velocity. Section 3.6.4 reports the method used for calculating closed-form SEP for correlated channel using the MGF of the channel fading statistic fot the Alamouti scheme. The proposed dynamic allocation with feedback is presented in Section 3.6.5 and results are shown in terms of BER gains.

3.6.2 System Model

OFDM is an excellent transmission technique for broadband wireless links since it divides the frequency-selective (time-dispersive) channel into a number of narrow subchannels that can be considered as flat-fading via an easy implementation with DFT. In OFDMA systems, the data to be transmitted is organized in frames which can be represented as a grid or a matrix with $N_{\text{data}} \times N_{\text{ofdm}}$ entries [114], where N_{data} is the number of data subcarrier per OFDM symbol and N_{ofdm} is the number of OFDM symbols in one frame.

[11] In order to have large bandwidths, next generation systems will need to exploit higher frequency bands thus worsening the Doppler effect which is directly proportional to the carrier frequency.

The number of data subchannels, N_{data} in which the spectrum is divided depends on the number of the IDFT/DFT points N_{fft}. The number of IDFT/DFT points are usually bounded to the powers of 2 for easy implementation with the radix-2 and radix-4 algorithms [24].

The total IDFT/DFT points are equal to sum of data subcarriers plus guards and pilots: $N_{\text{fft}} = N_{\text{data}} + N_{\text{guards}} + N_{\text{pilots}} + 1$ where 1 is the central DC tone.

If a system with N_T transmitting antennas is considered, N_T frames are composed, one for each antenna. The available subcarriers for data transmission are less than the total number of DFT points because part of the tones are used for guard intervals and then set to zero and some of them are assigned to pilots or training sequence transmission. The collection of all the subcarriers along frequency is called an OFDM symbol. T_{ofdm}, time duration of the OFDM symbol, is related to the inverse of the frequency subcarrier separation, ΔF. T_{ofdm} and ΔF are defined in the first phase of the system design [114]. The OFDM symbol duration T_{ofdm} and the subcarrier spacing in frequency ΔF are fundamental for calculating the STBC BER performance since they represent the minimum distance, thus maximum channel correlation, that the symbols can experience once allocated on the OFDMA grid [114].

Figure 3.16 shows a diagram of a MIMO-OFDM system with a generic STBC encoder. The symbols from the source are first modulated and then enter the STBC encoder. The number of the outputs of the encoder is equal to N_T, the number of transmitting antennas. For each of these output, a mapping function places the coded symbols on the frame based on a single-bit feedback from the MT.

3.6.3 The Time-Frequency Correlation in the Frequency Domain

A model of the correlation coefficients ρ_t and ρ_f between subcarriers both along time and frequency direction respectively is needed in order to asset a generalized criterion to decide in which direction the STBC symbols should be allocated. In Section 3.6.4, the SEP of the Alamouti scheme is computed for different correlation coefficients which are obtained based on the derivation of this section.

Almost in all the practical cases, the radio channel behavior can be effectively described as a Linear Time Variant (LTV) system. Stemming from the nomenclature used in [9], where a set of dual frequency-time functions were

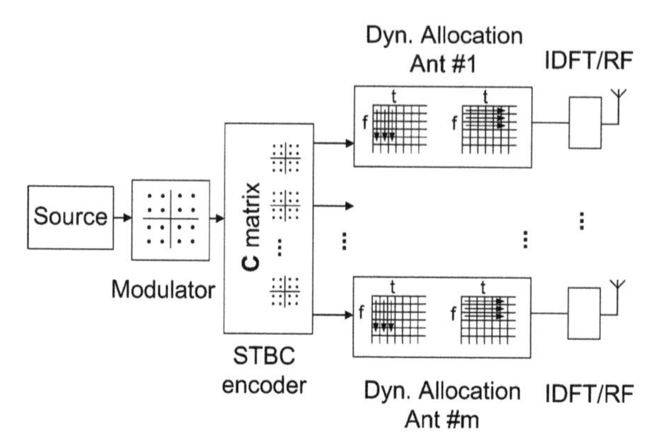

Figure 3.16 MIMO-OFDM system model with STBC encoder

defined for LTV systems, the performance of diversity schemes based on the *spaced-frequency, spaced-time correlation function*: $\phi_H(\Delta t, \Delta f)$ is studied.

In a multipath channel the *time-variant channel impulse response* $h(t, \tau)$ can be expressed as a sum of various paths:

$$h(\tau, t) = \sum_i \beta_i(t) e^{-j\theta_i(t)} \delta(\tau - \tau_i(t)) \tag{3.45}$$

where the index i represent the propagation paths, $\beta_i(t)$ are the time-varying path gains and $\theta_i(t)$ are the path phases. Each path is delayed, respect to a first reference path, of $\tau_i(t)$.

From the *time-variant channel impulse response*, $h(t, \tau)$, the *time-variant channel transfer function* in the Frequency Domain (FD), $H(f, t)$, can be expressed via the Fourier transform as

$$\begin{aligned} H(f, t) &= \mathcal{F}\{h(\tau, t)\} = \int_{-\infty}^{+\infty} h(\tau, t) e^{-j2\pi f \tau} d\tau \\ &= \sum_i \beta_i(t) e^{-j(2\pi f \tau + \theta_i(t))} \end{aligned} \tag{3.46}$$

The baseband equivalent of the channel matrix $\mathbf{H}(i, j)$ ($h_{ij} \in \mathbb{C}$) is obtained sampling the value of $H(f, t)$ at $f = f_0 + \Delta F \cdot i$ where f_0 is the lowest frequency subcarrier and at $t = T_0 + T_{\text{ofdm}} \cdot j$. In Figure 3.17 the channel magnitude ($\|\mathbf{H}_{ij}\| - [\text{dB}]$) of one frame in the frequency domain is reported. The channel model is the ITU Vehicular Channel Type A with a terminal velocity of 100 km/h. The plot shows the channel frequency profile

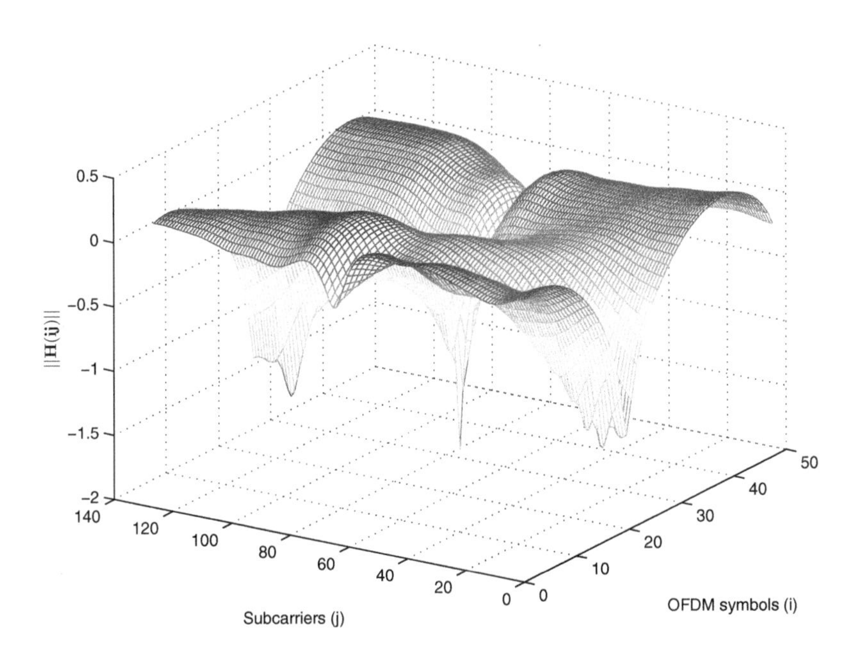

Figure 3.17 OFDM transfer function vehicular channel, 128 subcarriers and 48 OFDM symbols, 150 km/h

and its variation over time due to MT mobility. When MIMO is used, a two-dimensional channel transfer function is measured for each combination of two transmitting-receiving antennas.

The general *channel correlation function* in the FD is

$$\phi_H(f_1, f_2, t_1, t_2) = E\{H^\dagger(f_1, t_1)H(f_2, t_2)\} \tag{3.47}$$

Under the assumption that the channel is WSSUS, the correlation function is depending only on time and frequency differences, thus (3.47) can be written as

$$\phi_H(\Delta t, \Delta f) = E\{H^\dagger(f, t)H(f + \Delta f, t + \Delta t)\} \tag{3.48}$$

which is the *spaced-frequency, spaced-time correlation function* in the frequency domain.

The real value of $\phi_H(\Delta t, \Delta f)$, sampled at multiplies of ΔF and T_{ofdm}, represents the correlation between various $\mathbf{h}_{ij}$: $\phi_H(i, j)$. The correlation

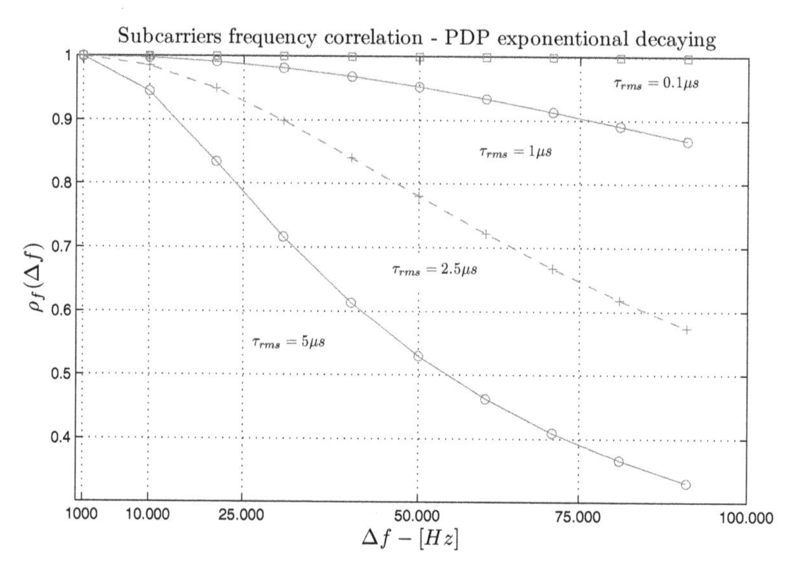

Figure 3.18 Exponential decaying PDP – subcarriers correlation over frequency separation for various τ_{rms} delays

along time or frequency direction on the OFDMA grid can be calculated as

$$\begin{aligned} \rho_f(i) &= \phi_H(0, \Delta f)|_{\Delta f = f_0 + i \cdot \Delta F} \\ \rho_t(j) &= \phi_H(\Delta t, 0)|_{\Delta t = T_0 + j \cdot T_{\text{ofdm}}} \end{aligned} \tag{3.49}$$

The minimum time and frequency separations are lower bounded by the OFDM system design parameters, respectively T_{ofdm} and ΔF.

The functions $\phi_H(\Delta f)$ and $\phi_H(\Delta t)$ can be obtained by evaluating the spaced-time spaced-frequency correlation function as in (3.49) by nulling the domain which is not of interest or also by evaluating the Fourier transforms of the *PDP* $\phi_h(\tau)$ and of the *Doppler spectrum* $S_H(\nu)$ (where ν is the Doppler shift variable) respectively. The proposed dynamic allocation can minimize the BER for the STBCs through the estimation of the PDP and of the Doppler spectrum. In OFDM systems, the RMS delay spread τ_{rms} and the Doppler frequency can be easily estimated and are usually calculated at the receiver. Various methods have been proposed in literature [92, 119, 121, 129]. When only the τ_{rms} and f_d are available, an assumption about the shape of the PDP and of the Doppler spectrum is needed.

Table 3.5 PDP models and $\phi_h(\Delta f)$ correlation function

Doppler Spectra	$S_H(\nu)$	$\phi_H(\Delta t)$
Jakes Model	$\begin{cases} \frac{1}{\pi f_d}\frac{1}{\sqrt{1-\left(\frac{f}{f_d}\right)^2}}, & \text{if } \lvert f \rvert < f_d \\ 0, & \text{otherwise} \end{cases}$	$J_0(2\pi f_d \tau)$
Two Rays	$\begin{cases} \frac{1}{2f_d}, & \text{if } \lvert f \rvert < f_d \\ 0, & \text{otherwise} \end{cases}$	$\text{sinc}(f_d \tau)$
Uniform	$\frac{1}{2}\left[\delta(f+f_d)+\delta(f-f_d)\right]$	$\cos(2\pi f_d \tau)$

If an exponential decaying PDP model is assumed, the related spaced-frequency correlation function can be expressed in function of the τ_{rms} as

$$\phi_H(\Delta f) = \frac{1}{1 + j2\pi\tau_{\text{rms}}\Delta f} \tag{3.50}$$

The frequency correlation (ρ_f) is illustrated in Figure 3.18 when the PDP is assumed as in (3.50); various correlation are reported for different channel delay spreads τ_{rms} and subcarriers spacing ΔF.

The correlation among successive OFDM symbols on the same subcarrier index is related to the MT velocity and the Doppler spectrum shape. The Doppler spectrum shape depends on the geometrical assumption made on the directions of the multipath signals impinging on the receiving antenna. The most common model assumed is the Jakes model [25] which is between the two extreme cases of two rays and uniform distribution of the arriving rays at the receiving antennas.

In Table 3.5, the three most common cases are reported with the respective space-time correlation functions.

The time correlation is shown for Jakes spectrum in Figure 3.19. Figures 3.20 and 3.21 plot the theoretic correlation among subcarriers for the other two Doppler models. The Doppler frequencies shown are corresponding to MT velocities of 60–125–250 km/h at a carrier frequency of 6 GHz. In the case of Jakes model, when $T_{\text{ofdm}} = 100\ \mu\text{s}$, the correlation coefficient ranges from 1 to 0.8. The Jakes model represent an of average situation for the impinging signals on the antenna that can be encountered in typical urban scenario. However, when the MT is in a worse scenario and the channel has dominant multipath rays, meaning that energy is received at the MT antenna concentrated from certain directions, the correlation can be much lower compared to the Jakes model. In the case of two rays model, it ranges from 1 to almost 0.6, as visible in Figure 3.20. The uniform model is providing the

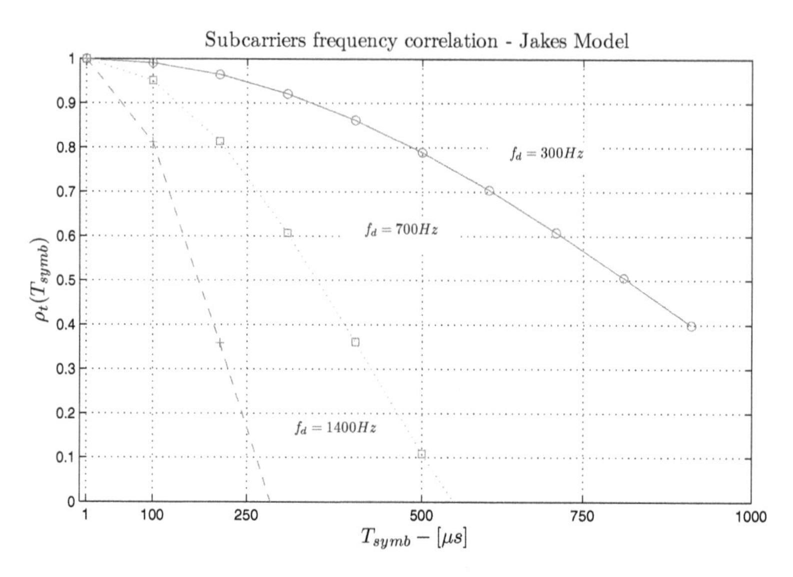

Figure 3.19 Jakes model – subcarriers correlation over time for various Doppler frequencies

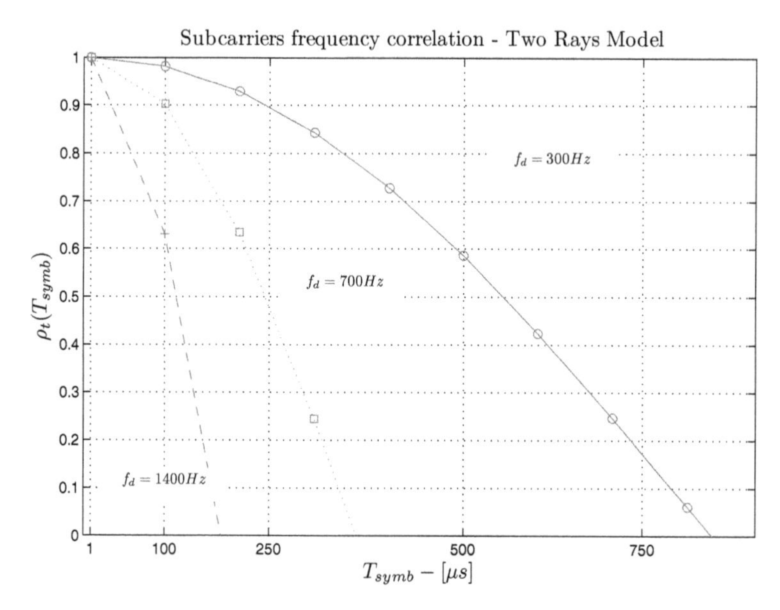

Figure 3.20 Two Rays model – subcarriers correlation over time for various Doppler frequencies

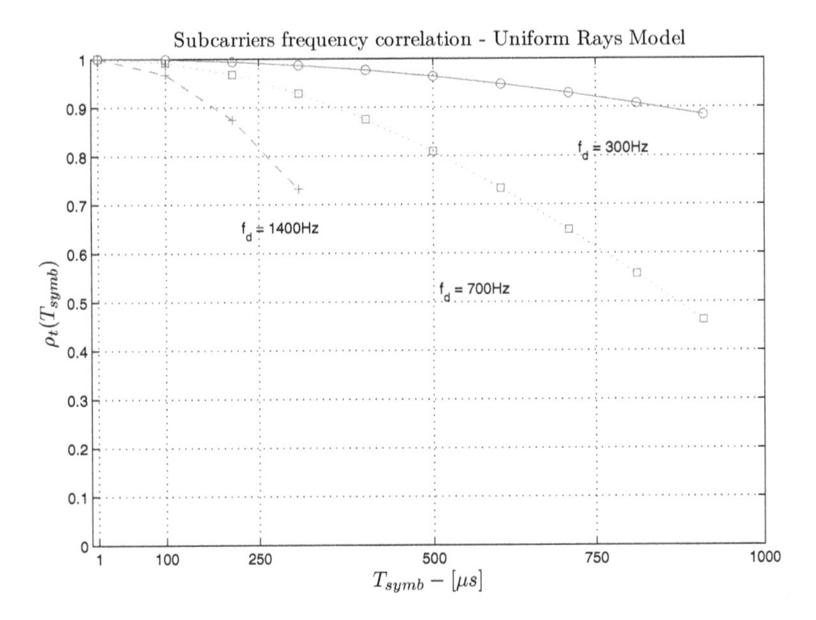

Figure 3.21 Uniform model – subcarriers correlation over time for various Doppler frequencies

highest correlation values but represents the ideal situation of a very high scattering channel where the antenna at the MT sees signals coming from all the direction in the space. Further, high scattering channels are usually encountered indoor which is not the typical environments for medium and high velocity MTs.

Comparing the values of ρ_t and ρ_f from Figures 3.18 and 3.19, we note that the correlation coefficient can be lower in the time or frequency domain, depending on the relative ratio of terminal velocity and delay spread. Thus, the BER performance of STBCs can be enhanced if a dynamic allocation direction is implemented as shown in Section 3.6.5.

3.6.4 Closed-Form SEP Calculation in Uncorrelated Channels

A space-time block code is defined by a coding matrix which maps constellation symbols over the N_T transmitting antennas in successive T time slots. The coding matrix $\mathbf{A}$ has to be orthogonal since a non-orthogonal $\mathbf{A}$ will result in ISI interference at the receiver, as explained in Section 3.2.1. Matrix $\mathbf{A} \in \mathbb{R}^{L \times N_T}$ maps $L \times N_T$ data symbols over successive L time slots over the transmitting antennas. The decoding process for the ST codes is simple and it is based on linear combination of the transmitted signals from each of the N_T transmitting antennas.

The Alamouti coding matrix **A** can be written as (see Section 3.2.1)

$$\mathbf{A} = \begin{bmatrix} s_1 & s_2 \\ -s_2^* & s_1^* \end{bmatrix} \tag{3.51}$$

STBCs assume that the channel remains constant during the time duration of the code (L signalling time slots). Alamouti code requires that the channel is constant over 2 signalling period. The constant channel assumption is easily not verified particularly in quick time variant channels.

When channel coefficients are not correlated ($\rho < 1$), the BER performance of the STBCs is influenced by the correlation coefficients between two consecutive symbols: ρ_f if symbols are allocated along frequency, ρ_t otherwise.

The SEP performance for the Alamouti transmit diversity scheme with a time-varying channel has been derived for narrow-band partially correlated Rayleigh channels for BPSK, in [115]. We can express the channel coefficients accounting for the correlation as

$$h_1^1 = \rho h_1^2 + \sqrt{1-\rho^2}\epsilon_1 \tag{3.52}$$

$$h_2^2 = \rho h_2^1 + \sqrt{1-\rho^2}\epsilon_2 \tag{3.53}$$

where h_j^i is the complex channel coefficient, j is the transmitting antenna and i is the time slot of the Alamouti block, $i|j \in \{1,2\}$. The random variables ϵ_i are distributed with Rayleigh distribution, as the channel coefficients. If the correlation $\rho = 1$, the transmit diversity scheme is not suffering any performance losses. The other extreme case is represented by $\rho = 0$, when no correlation exists between two successive channel coefficients.

When ρ is comprised between these two extremes, the SEP for a Zero Forcing (ZF) detector can be expressed based on the SNR distribution at the output of the STBC combiner. In [115], the closed-form SEP for BPSK has been derived. The calculations are derived here for QAM modulation of any order as in [81].

If the channel coefficient h_j^i are assumed to be Rayleigh distributed, the SNR distribution at the output of the receiver is

$$p_\gamma(\gamma) = \frac{2(1-\rho^2)}{\overline{\gamma}} e^{-2\gamma/\overline{\gamma}} + 4\rho^2 \frac{\gamma}{\overline{\gamma}} e^{-2\gamma/\overline{\gamma}} \tag{3.54}$$

To obtain the averaged SEP over the fading channel, the SEP conditioned on the instantaneous SNR needs to be integrated over the SNR distribution.

$$\mathrm{SEP}_M(\overline{\gamma}) = \int_0^{\infty} \mathrm{SEP}(M, \gamma) p_\gamma(\gamma) d\gamma \tag{3.55}$$

where $\overline{\gamma}$ is the average SNR. Following the approach in [98], the SEP for coherent detection of Alamouti symbol with QAM modulation of any order is derived as

$$\mathrm{SEP}_M(\overline{\gamma}) = 4KI^{(1)}(a(M), \gamma) - 4K^2 I^{(2)}(a(M), \gamma) \tag{3.56}$$

where $K = (\sqrt{M} - 1)/\sqrt{M}$, M is the modulation order and

$$a(M) = \sqrt{(3\log_2(M)/(M-1))}$$

$$I^{(1)}(a(M), \gamma) = 1/\pi \int_0^{\pi/2} M_\gamma(-s(\theta)) d\theta$$

$$I^{(2)}(a(M), \gamma) = 1/\pi \int_0^{\pi/4} M_\gamma(-s(\theta)) d\theta$$

In the integrand functions: $s = a^2/2\sin^2(\theta)$ and $M_\gamma(s)$ is the Moment Generating Function (MGF) function of the SNR distribution at the output of the ZF receiver:

$$M_\gamma(s) = \frac{(1-\rho^2)}{(s\overline{\gamma}/2+1)} + \frac{\rho^2}{(s\overline{\gamma}/2+1)^2} \tag{3.57}$$

In Figure 3.22, analytical SEP curves are reported for 4–16–64-QAM modulations with correlation coefficients $\rho \in [1, 0.9, 08]$. For SNR levels above 4 dB, the Alamouti scheme is very sensitive to channel variability and the performance is highly degraded. For a 4-QAM modulation at SNR = 20 dB, the SEP shifts from around 10^{-4} to more than $3 \cdot 10^{-3}$.

3.6.5 Performance Analysis

In Section 3.6.3, the correlation coefficient along time and frequency allocation direction have been analyzed, based on the *spaced-time, spaced-frequency correlation function* in the frequency domain for a WSSUS channel. The SEP has been expressed in closed-form for the Alamouti scheme in Section 3.6.4.

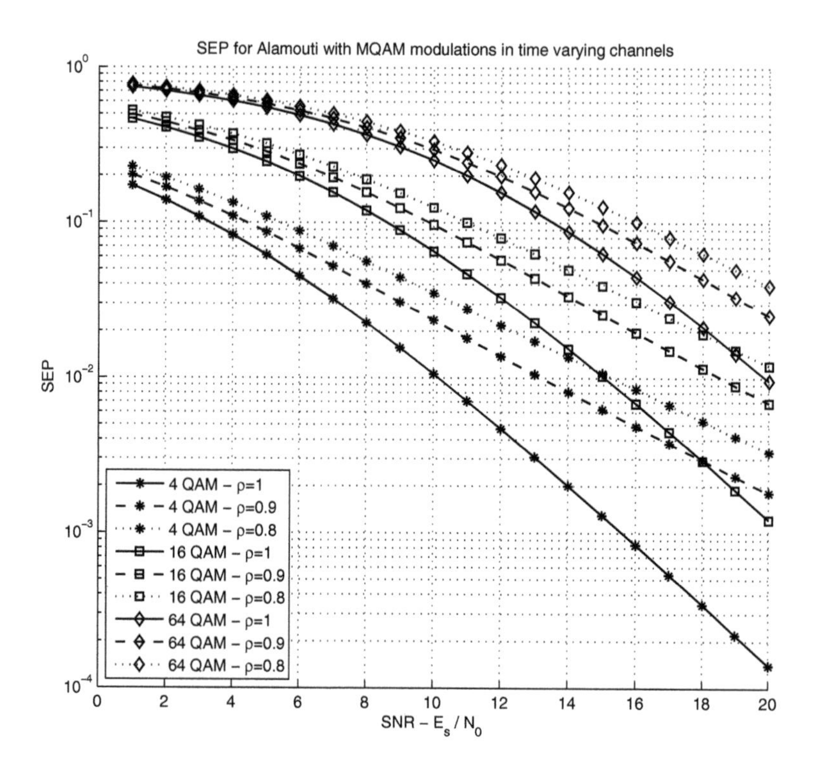

Figure 3.22 SEP for Alamouti scheme for 4–16–64-QAM modulations over narrow-band partially correlated Rayleigh channel

Table 3.6 STFBC: system parameters

Parameters	Description	Value
c_f	Carrier frequency	2.5 GHz
BW	Total bandwidth	10 MHz
N_{fft}	Number of FFT points	1024
ΔF	Subcarrier spacing	10.9375 kHz
T_{ofdm}	OFDM symbol duration without CP	91.43 us
CP	Cyclic prefix length	1/8 T_{ofdm}
$T_{\text{ofdm}}^{\text{CP}}$	symbol duration with CP	102.86 us
F	Frame length	5 ms
N	Number of OFDM symbols in frame	47

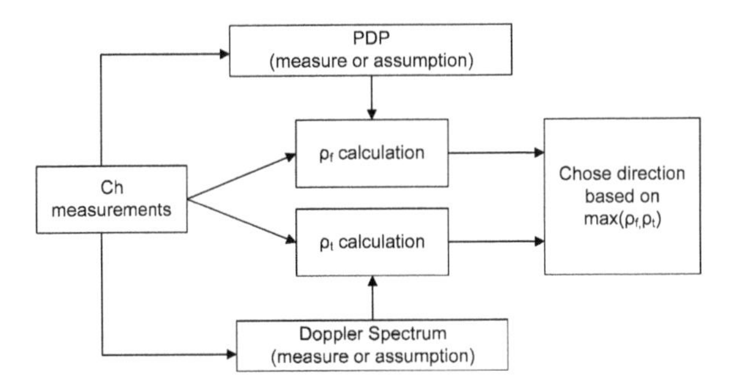

Figure 3.23 The dynamic allocation direction of STBC in an OFDM system

The adaptive allocation on the OFDM grid based on the estimated delay spread τ_{rms} and Doppler frequency f_d is here proposed. In Table 3.6 we report the OFDM system settings, referring to the standard IEEE802.16e.

The proposed system is illustrated in Figure 3.23. The system calculates ρ_t and ρ_f. Then it determines the allocation direction where the correlation is higher. These two coefficients can be exactly calculated if the PDP and the Doppler spectrum of the channel are known. Assumptions can be made on the Doppler spectrum and PDP shapes. In that case, the system needs only to estimate the channel delay spread and the Doppler frequency.

The computer simulations show the Symbol Error Rate (SER) performance of the OFDM system with the Alamouti scheme in a DL connection to a high velocity MT moving at 150 km/h. The results are averaged over a simulated real-time of 10 seconds. The dynamic allocation operates with the assumption of Jakes Doppler spectrum and Exponential decaying PDP. The ITU vehicular channel model A has a $\tau_{\text{rms}} = 3.32\ \mu$s. In the case of the exponential decaying PDP, we can calculate the expected correlation between adjacent subcarriers along the frequency direction obtaining $\rho_f = \phi_H(\Delta F) = 0.98$, $\Delta F = 10.9375$ kHz.

In Figure 3.24, we can see that the dynamic allocation direction preserves the diversity gain of the Alamouti scheme. The time allocated STBC is losing performance due to low correlation between symbols allocated on the same subcarrier but on different time slots. The dynamic allocation is able to track the best correlation and dynamically change the allocation direction of STBC. This reduces the BER and preserves the system throughput.

Since the MT velocity is high (150 km/h), the best correlation is over the frequency domain ($\rho_t < \rho_f$). The SER of the dynamic allocation tends to the theoretical one calculated in closed-form with the frequency correlation

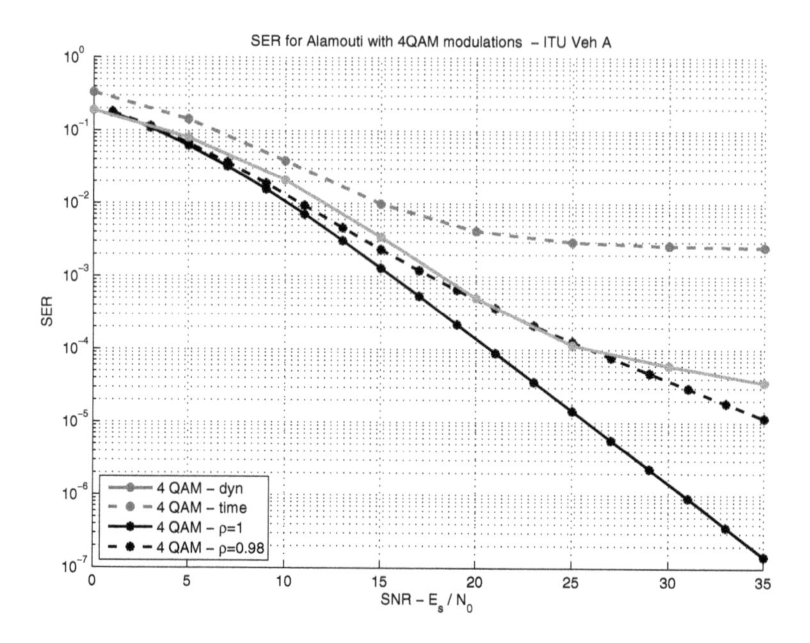

Figure 3.24 BER improvement for ITU vehicular channel model A with MT at 150 km/h

coefficient $\rho_f = \phi_H(\Delta F)$. The difference between the theoretically calculated Alamouti BER performance over correlated channels and the simulated one derives from the fact that the closed-form calculation is valid for narrowband transmissions. The simulation results reported are the average of STBC transmitted symbols over all the subcarriers of the frame and include the effect of ICI coming from Doppler shift.

In the case of receivers with very low Doppler frequencies,[12] the symbols are likely to be mainly allocated along the time direction, especially in scenarios with high delay spreads, providing better SER performance respect to a fixed allocation along the frequency direction.

The multipath channel has been simulated with Rayleigh statistic distribution. Other fading statistic model can be inserted in the analysis. The presence of various degrees of LoS can be studied using Nakagami-m fading statistics. Correlated space-time coding over correlated Nakagami-m channel fading

[12] This can be the case of static MTs which generally show an f_d of the order of some units of Hertz, due to moving object in the radio environment.

has been studied in [73, 97]. The Nakagami index m has an important role being directly related to the channel correlation in the frequency domain.

Based on these derivations, multivariate Gamma distribution can be used to described the SNR at the receiver thus obtaining SEP for the correlated STCs over Nakagami-m channel. When the fading severity factor of the channel is decreasing (when $m > 1$ in the Nakagami-m channel statistic), the channel will show a reduced frequency selectivity, raising the correlation between adjacent subcarriers: when LoS component is sufficiently high, SFBC would perform better also with medium to high speed MTs.

3.7 Conclusions

In this chapter, MIMO has been studied aiming at PHY layer enhancements. The performance gains are coming from the added degrees of freedom of the radio signals which can be exploited with suitable receiving techniques. This approach can be fragile and the performance are very sensitive to several physical characteristics of the channel and of the devices, among them:

- antenna correlation at transmitter and receiver,
- MIMO channel matrix rank,
- channel time variability.

The impact of MIMO channel matrix rank and channel time variability have been studied. Two solution have been proposed to dynamically adapt to the radio channel dynamics: an adaptive MIMO scheme selection, based on upper-bounded BER and an adaptive STFBC transmission for systems based on OFDM/OFDMA.

Hybrid LSTBC scheme have been analyzed jointly and SM have been analyzed for a system constrained with a multiplexing gain of 2. If the diversity constrain is not required, Alamouti scheme has been used instead of LSTBC in order not to suffer from BER degradation due to the quasi-orthogonal LSTBC structure.

A solution with a light feedback channel has been proposed for fast channel variation in the presence of high velocity MTs via a dynamic STFBC allocation over the OFDMA frame. The system proposed is able to adapt the allocation direction on the OFDMA grid in order to best match to the channel characteristics of the MT. Current standards (as IEEE802.16) are using fixed allocation direction and they may lose the diversity enhancement of the STBC. The proposed dynamic allocation requires low modifications in the air-interface design and preserves the STBC performance which otherwise would be lost due to the MT mobility. The feedback needs only a single-bit feedback.

The strategy has been simulated with the Alamouti scheme, without loss of generality. The results can be extended to a general L length STBC where the improvements will show even greater enhancements, considering that the assumption of quasi-static channel is more difficult to be verified when the STBC length increases.

4

MAC Layer: ARQ and DSA Analysis

4.1 Introduction

This chapter is focused on some of the mechanisms at the MAC layer of the IEEE802.16 stack and its three sublayers: the security sublayer, the Link Layer Control (LLC) sublayer and the MAC sublayer. Usually, the functionalities of these three sublayers are kept separate from PHY layer; however, some CL strategies have been proposed, as in [82], where a subcarrier permutation at the PHY layer is used to encapsulate higher layers' security mechanisms to provide enhanced network security.

At the LLC, the communications are organized in flows and are managed with a connection-oriented approach. Each service flow has a connection-ID as specified by the reference standard for LLC: IEEE802.2 [1].

The MAC sublayer manages the ARQ retransmissions and the multiple access to the radio resource. In 802.16 the multiple access to the channel is supported with dynamic resource allocation. Depending on the PHY layer configuration – SC, SCa, OFDM, OFDMA – the dynamic resource allocation follows different approaches. In OFDMA, the DL and UL frames are divided in chunks: a set of subcarriers that can be assigned to a service flow connection. The resources are allocated both in the time and frequency domain via a two-dimensional partition of the OFDMA grid, which is visible in Figure 3.17.

The resource assignment mechanism in OFDMA is based on statistical multiplexing, which is a multiplexing technique where the communication channel (in this case, the RF bandwidth of the cell) is partitioned into a number of variable-capacity digital channels. The resources are shared and the allocation is adapted to the instantaneous traffic demands of the service flows that are admitted in the system. Statistical multiplexing is opposed to the fixed multiplexing where the sharing of the resource is static, as in TDM.

Statistical multiplexing provides a system performance gain due to the opportunistic scheduling of the admitted service flow. This gain is called *statistical multiplexing gain* and permits higher link utilization factor: the scarce resources of the wireless networks (namely time and spectrum) are thus better utilized. Statistical multiplexing accounts for the QoS requirements of each service flow when the priorities for the access to the radio resources are calculated; for instance, when the priorities based on rate and delay measurements at the scheduler queues as in [80].

Statistical multiplexing with QoS awareness and aggressive resource exploitation is facilitated by a packet-based design. The packet segmentation of the traffic flows permits better flexibility and network performance.[1]

In this chapter the following topics will be analyzed:

- MAC sublayer: PHY/MAC CL optimization for a Stop and Wait ARQ protocol
- LLC sublayer: Dynamic Service Addition blocking analysis.

4.2 ARQ: A Cross-Layer Strategy for QoS-Guaranteed Links

Future wireless networks are going to offer high transmission rates to stationary and mobile users supporting the convergence of data, voice, and video transmissions. Such convergence imposes strict requirements on the QoS that must be ensured to users' transmission. The problem of ensuring QoS to 4G network users is especially hard due to the high data rate involved and the high mobility of the MTs. QoS guarantees are extremely difficult to maintain in wireless networks, especially due to the intermittent quality degradation that radio channel may experience. Degradations and fluctuations of the channel quality can be caused by various impairments and are further exacerbated by high user mobility and high data rates. In order to achieve the requested QoS, strategies that counteract the channel quality variations in an *adaptive* way are necessary.

Aggressive QoS strategies can be greatly enhanced via cross-layering design. The classic layered communication stack has been shown to be very useful for a coordinated technological development of data networks. How-

[1] The more complicated prediction of the network load and the more complex admission control mechanisms are two drawbacks of packet-based statistical multiplexing compared to the circuit-switched designs. In circuit-switched networks, once the line has been leased for a communication, QoS is met and the load is readily calculable.

ever, dividing the problem of communication in strictly separated layers has reduced the overall network efficiency. It is true that some operations of the data networks are fundamentally separated in their nature and cannot be solved with a single approach. In this sense, the study of physical communications (radio, optic, cable, etc.) are a widely separated research fields from those studying networking capabilities with data flowing from multiple sources to multiple destinations. However, strictly layered stack, even if represents a good base for standardized and well-organized technical development, greatly reduces the performance and the flexibility of a wireless network. For a good overview of the fundamental capabilities of a data network and the limitations in the current layered stack, the reader is referred to [88].

An Adaptive Cross Layer (ACL) strategy that jointly optimizes the parameters of PHY and MAC layers using queueing theory is proposed and applied to a mobile user communication in an IEEE802.16 wireless network, supporting different modulations and the ARQ protocol. PHY and MAC layer parameters are optimally and dynamically combined by ACL strategy to meet the QoS requirements.

The evaluation of the impact of mobility on ACL strategy performance is assessed by using an accurate channel model that accounts for fast fading and shadowing effects. The simulation results show that the ACL strategy outperforms non-adaptive or single-layer strategies, in terms of bandwidth savings. The ACL strategy is able to guarantee the requested QoS in the different mobility scenarios.

4.2.1 State of the Art in Cross-Layer ARQ

Adaptive strategies for guaranteeing QoS may exploit the various degrees of flexibility and redundancy that are offered by the different layers of a 4G wireless network. It is well known that the choice of the layer at which the adaptive strategy is implemented affects the overall performance of the wireless network. In particular, strategies implemented at the upper layers (e.g., IP layer) could slowly counteract the channel variations. In addition, if different disjoint strategies are implemented at the different layers, lower layers (e.g., the PHY layer) may tend to promptly under- or overcompensate the channel quality degradation, and thus the upper layer may strive to achieve the QoS requirements. Therefore, the adaptive strategy must feature a cross-layer approach to optimally achieve the QoS requirements and a prompt adaptivity to compensate channel quality variations.

A number of previous works considered cross-layer strategies suitable for 4G networks (e.g., [70, 117]). The performance of the proposed solutions has been mainly evaluated for stationary users. The effects of slowly varying Channel Quality Indicator (CQI) on ARQ protocol are evaluated in [6]. In such work, channel state information is derived as long-term BER. The estimation is computed on a sliding window and four simple adaptation algorithms are provided. These strategies can react only after a certain number of errors, thus with a delay that is not tolerable in real time traffic.

A more accurate channel model that accounts for CQI variations is considered in [69, 70, 94, 117]. The model is based on a Finite State Markov Chain (FSMC) that alternates good and bad channel quality states, according to simulation parameters. Transition probabilities and state holding times are inferred from statistical measurements of the wireless channel and lack of a proper model for the user mobility inside a wireless network cell.

4.2.2 Adaptive CL ARQ System Model

The proposed ACL strategy jointly and promptly optimizes the PHY and MAC parameters [14]. The objective of ACL strategy is twofold, i.e., to meet QoS requirements and to allocate the minimum amount of bandwidth necessary for transmissions. The proposed adaptive strategy is applied to a IEEE802.16e network based on an OFDM PHY layer; system parameters are set according to [15, 47, 114]. The PHY and the MAC layer of IEEE802.16 support, respectively, different modulation profiles and a stop-and-wait ARQ protocol. They are optimally combined by the ACL strategy. The optimization is based on the analytical evaluation of the QoS performance in terms of packet loss and expected delay, derived using queueing theory.

The ACL strategy is applied to a mobile user scenario, in which the transmission channel is affected by both fast and slow fading. The slow-fading inter-frame process is originated by the shadowing due to buildings or obstacles encountered along the mobile user trajectories. The performance of the ACL strategy is evaluated assuming perfect and immediate knowledge of CQI. Simulation results assess the adaptivity and the optimality of the ACL strategy under different user mobility scenarios and quantify the impact of user mobility on QoS.

The ACL strategy is tested over the radio link between the BS and a single MT. BS and MT communicate using IEEE 802.16 protocols operating in point-to-multipoint mode.

IEEE 802.16 is a connection-oriented protocol and transmissions are time-division multiplexed. In each time frame, the BS schedules the transmissions and indicates the DL and UL bandwidth allocation in a map along with the PHY layer parameters. Data packets, i.e., MAC PDU, are transmitted during the scheduled time allocation.

Without loss of generality, a single connection, referred to as "service flow", is established on the radio link from BS to MT. The following assumptions are made on the derivation of the CL layer optimization scheme.

Further details regarding IEEE802.16 packet processing can be found in Chapter 6.

It is assumed that the transmitted power at BS is kept constant. The BS has the perfect and immediate knowledge of the CQI. The service flow is an Unsolicited Grant Service (UGS) type. The PDU packets of UGS have fixed size and are assumed to be periodically issued (e.g., UGS supporting voice traffic streams). Issued PDUs are stored in a buffer with unlimited capacity. In each time frame, the bandwidth required for PDU transmission is dynamically and optimally allocated upon ACL strategy computation.

4.2.3 ACL Strategy

The main objective of the cross-layer optimization is to minimize the amount of bandwidth necessary for transmitting the service flow PDUs, while ensuring the requested QoS. The bandwidth required to transmit a single PDU is defined as the bandwidth (e.g., allocated time) for the first time transmission of the PDU and for the subsequent retransmission of the same PDU, until either the PDU is successfully received or the maximum number of retransmissions is reached.

QoS requirements include maximum PDU loss rate, $L_{\max}$, and may include a set of other QoS requirements (such as delay and jitter); the QoS set is indicated with Q.

To minimize the bandwidth while ensuring QoS requirements, the PHY and MAC layer parameters are optimally and jointly decided. The PHY layer parameters (e.g., modulation order, code type, and rate) allows to define the PHY profile to select. The MAC layer parameters include whether ARQ protocol is required, and if so the maximum number of times, K, the transmission of the same PDU should be attempted, before dropping the PDU.

Let S be the set containing the SE of each PHY profile. Let λ be the expected arrival rate of PDUs. Let P_i be the expected probability that a fixed-size PDU, transmitted using PHY profile i, is received incorrectly, for a given

CQI measured at the PHY layer. The expected P_i can be calculated measuring the channel status and using a look-up table when complex MCS are used and no closed-form expressions are available.

Under the assumption of unlimited buffer capacity, PDU losses occur only when the maximum number of transmission attempts (i.e., K) is reached. Thus, let $f(L_{\max}, P_i)$ be the function that optimally selects the MAC parameter K to meet $L_{\max}$ values. Also, assume to know the set of functions $g_j(K, \lambda, P_i)$ that evaluate each QoS metric j.

The CL optimization works as follows. The various PHY profiles are considered, starting from those at maximum spectral efficiency; a set of physical layer profile with the same SEs (which is the maximum possible) is found as

$$\xi = \max_i S[i] \tag{4.1}$$

Among the PHY profiles having the same spectral efficiency, the PHY profile at minimum PDU error rate is selected:

$$\rho = \min_{i:S[i]=\xi} P_i \tag{4.2}$$

Then, the optimal number of transmission attempts, K_{opt}, for the given value of ρ and $L_{\max}$ can be derived as

$$K_{\text{opt}} = f(L_{\max}, \rho) \tag{4.3}$$

If constraint $g_j(K_{\text{opt}}, \lambda, \rho) < Q[j]$ holds for each QoS metric j, then the optimum MAC parameter K_{opt} and PHY profile $s_{\text{opt}} = s$ are found. Otherwise, the QoS for the other PHY burst options at the same spectral efficiency ξ (if any) is evaluated. If QoS is not met, the other PHY profile options at decreasing SEs are evaluated.

Such CL strategy is performed each time the CQI deviates significantly. A flow-chart of the strategy is sketched in Figure 4.1. Notice that the PHY profile at maximum spectral efficiency is selected when QoS requirements are met. Also, for the same spectral efficiency, PHY profile with the best performance in terms of PDU error rate is selected.

A derivation of functions $f(\cdot)$ and $g_j(\cdot)$ is presented next.

The above presented ACL strategy is applied to the case in which the QoS requirements are given in terms of maximum PDU loss rate, $L_{\max}$, and expected PDU delay, $\overline{D}_{\max}$. The estimation of PDU loss rate, L, and the expected PDU delay, $\overline{D}$, are analytically derived in this section using a queueing model based on the following assumptions.

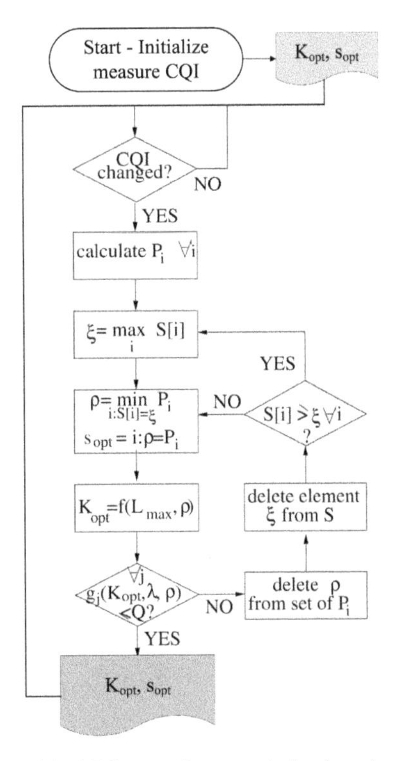

Figure 4.1 ARQ cross-layer optimization algorithm

Realization of the statistical process representing the BS-MT channel behaviour during a time frame are assumed to be independent on a time frame basis, i.e., fast fading is assumed to be block fading. BS can schedule a single PDU per time frame. Fragmentation or packing of PDU is not considered. ARQ acknowledgements are promptly received. Therefore, the ARQ protocol is based on stop-and-wait mechanism, with up to K transmissions of the same PDU, as shown in Figure 4.2.

Under the above mentioned assumptions, the performance of stop-and-wait protocol can be evaluated using a discrete time queueing model with Bernoulli arrivals at rate λ. Traffic model is Poisson distribution which is well-accepted for modelling voice calls and UGS service flows.[2] Let X be the expected service time; with Bernoulli arrival rates, first moment ($\overline{X}$) and

[2] More accurate traffic models can be used for (non-)real-time traffic: they may affect Admission Control (AC) blocking,but not the signalling blocking or delay.

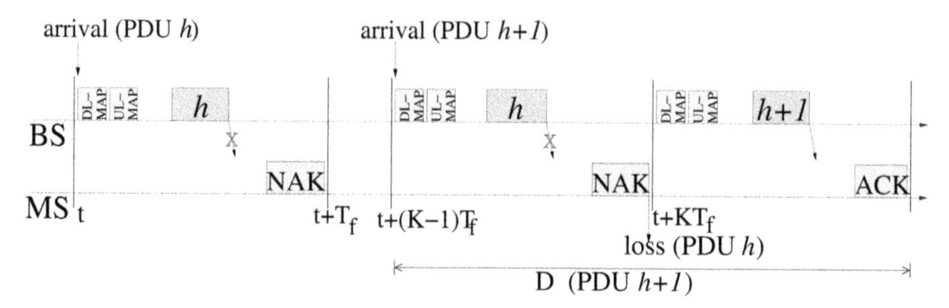

Figure 4.2 Considered stop-and-wait ARQ protocol

second moment ($\overline{X^2}$) of X are as follows:

$$\begin{aligned}\overline{X} &= T_f\left(\sum_{h=1}^{K} hP_i^{h-1}(1-P_i) + K\cdot P_i^K\right)\\ &= T_f\left(\frac{(1-P_i^{K+1})}{(1-P)} - (K+1)P_i^K\right) + T_f\cdot K\cdot P_i^K\end{aligned} \tag{4.4}$$

$$\begin{aligned}\overline{X^2} &= \left(\sum_{h=1}^{K} h^2P_i^{h-1}(1-P_i) + K^2\cdot P_i^K\right)\\ &= \left(\frac{(1+P_i - P_i^K(K+1)^2 + P^{K+1}(2K^2+2K-1)}{(1-P)^2}\right.\\ &\quad\left.+\frac{P_i^{K+2}\cdot(-K^2))}{(1-P)^2}\right)T_f^2 + T_f^2\cdot K\cdot P_i^K\end{aligned} \tag{4.5}$$

where T_f is the time frame duration.

The expected delay experienced by correctly received PDUs is given by the expected time spent in the buffer ($\overline{D_w}$) and the expected time spent for retransmissions (i.e., $\overline{X}$). The expected value of the first term, $\overline{D_w}$, is derived using the P-K formula [58]

$$\overline{D_w} = \frac{\lambda(\overline{X^2}-\overline{X})}{2(1-\lambda\overline{X})}, \qquad \overline{D} = \overline{D_w} + \frac{\overline{X}-K\cdot L}{1-L} \tag{4.6}$$

The PDU loss rate is given by $P_{i,n}^K$. In order to meet the PDU loss rate requirement, it is possible to derive

$$f(L_{\max}, P_i) = \left\lceil \frac{\log L_{\max}}{\log P_i} \right\rceil \tag{4.7}$$

and find K_{opt} as indicated in (4.3). Notice that such value of K_{opt} is the minimum value that allows to meet PDU loss rate requirement. Therefore, the bandwidth required to transmit a PDU is minimal. The QoS metric function concerning the expected delay is $g_j(K_{\text{opt}}, \lambda, P_i) = \overline{D}$, as in (4.6).

Finally, the expected amount of bandwidth necessary in each time frame for supporting the service flow is equal to $\lambda \cdot \overline{X}$ times the bandwidth required for transmitting one PDU.

4.2.4 Performance Analysis

In this section, after a detailed description of the channel model for mobile users and of the system parameters is provided, t the performance of the proposed ACL strategy is evaluated in different mobility scenarios. The ACL strategy is compared against two strategies. The first, a single-layer (SL) strategy, optimizes the MAC parameters only, according to the updated CQI, when PHY parameters are fixed and given. A second a cross-layer (CL) strategy, that is not aware of CQI variations over time, optimizes jointly PHY and MAC parameters, according to the average average channel quality (i.e., SNR).

The channel model accounts for mobility. The MT is assumed to move inside a network cell, along a circular path centred at the BS. The channel model for the MT is based on ITU Vehicular type A model [100]. MT mobility induces shadowing effects, that make the received average signal-to-noise, Γ, fluctuate according to a log-normal distribution. Path loss variations are neglected in this work as the changes are slower than shadowing, especially in urban mobility scenario. The ACL strategy which is able to react to the shadowing, can easily adjust the MCS to the slower variations of the link budget due to the long-term variations of the path loss.

For comparison purposes, a frequency flat fading model (LoS transmissions without Doppler effects) is considered. In the LoS case, power of the delayed multipath replicas is very low and no frequency selectivity is appreciated.

4.2.4.1 System Parameters

The performance of the proposed CL approach is evaluated for an OFDM-based PHY layer, according to the parameters indicated in Table 4.1. Parameters are defined according to IEEE 802.16 [47] standard. PHY profiles consist of three different types of modulation, i.e., 4-QAM, 16-QAM, and 64-QAM, applied to uncoded data. Carrier frequency is centered at 3.5 GHz

Table 4.1 ARQ ACL strategy, system parameters

Parameter	Value
Data subcarriers	384
FFT size	512
Subcarrier freq. separation	31.25 kHz
OFDM symbols in frame	48
Frame duration (T_f)	5 ms
Symbol duration	102.86 μs
Useful Symbol duration	91.43 μs
Cyclic Prefix Ratio	1/8

Table 4.2 ARQ ACL strategy, mobility scenarios

Scenario	Log-normal σ
Urban micro cell	2.3 dB
LOS fixed station	3.4 dB
Urban macro cell	8.0 dB
Indoor small office	12 dB

and the wavelength is 8.6 cm. MT speed is 50 km/h, which leads to a Doppler frequency of about 160 Hz for the specified carrier frequency. Shadowing is supposed to have a spatial coherence of 20 wavelengths. For the considered MT speed and time frame duration, shadowing causes a log-normal Γ deviation every 120 ms or 24 time frames [36, 39]. It is assumed that only 2 dB variations of Γ are detected (i.e., the log-normal distribution is quantized with a minimum step of 2 dB step over a given Γ range). The standard deviation of Γ depends on the mobility scenario. Shadowing parameters for the different mobility scenarios are selected according to Fu et al. [32] and reported in Table 4.2.

Upon detection of a channel quality variation, BS promptly performs the CL optimization, presented in Section 4.2.3, and schedules PDU transmissions according to the computed MAC and PHY parameters. The size of PDU packets, including MAC header and CRC, is fixed to be 288 bytes, i.e., 1 (2 or 3) OFDM symbol(s) at 64-QAM (16 or 4 QAM).

4.2.4.2 Numerical Results

In the simulation results, unless otherwise indicated, the ITU vehicular A channel model with log-normal deviation $\sigma = 2.3$ dB and Γ range in [26–40] dB is used. Unless otherwise stated, the QoS requirements to be met for any value of Γ are $L_{\max} = 10^{-4}$ and $\overline{D}_{\max} = 25$ ms. PDU arrival rate is $\lambda = 0.1$ PDUs per time frame. In the simulation, retransmissions of the

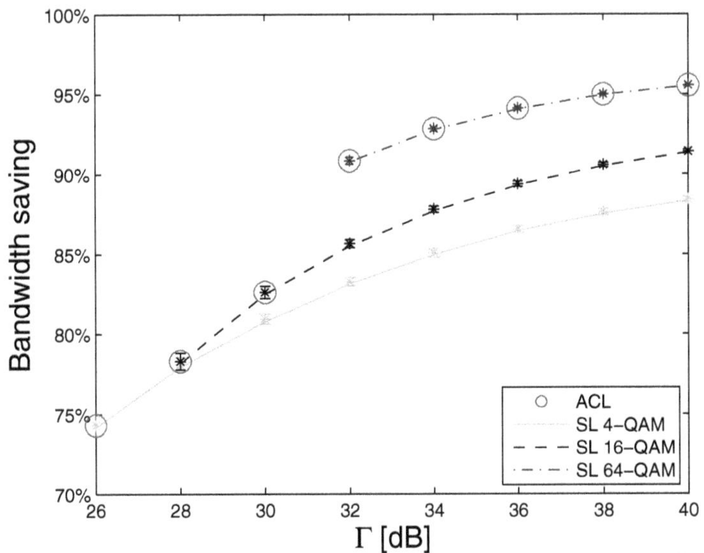

Figure 4.3 ITU vehicular A channel: percentage of bandwidth saving versus Γ, for ACL and SL strategies, $\sigma = 0$ (no shadowing)

same PDU use the same MAC and PHY parameters computed for the first transmission.

Figure 4.3 plots the percentage of bandwidth saved for transmitting PDUs of a single service flow (with QoS requirements) when bandwidth is dynamically allocated, instead of a fixed bandwidth allocation of three OFDM symbols per time frame. Dynamic bandwidth allocation is achieved using ACL strategy or an adaptive MAC (SL) strategy that optimizes MAC parameter K for a given modulation. Results are shown for different values of Γ, when $\sigma = 0$. In the figure, the curves are obtained using the queueing model, while the points with the error bars indicate the simulation results. Notice that at low values of Γ the modulations at higher spectral efficiency are unable to meet QoS requirements. The plot in Figure 4.3 shows the perfect matching between theoretical results and simulation results and the optimality of ACL approach. Moreover, the results quantify the bandwidth savings achieved by strategies with dynamic bandwidth allocation. Also, the results clearly indicate the best performance of the cross-layer strategy (i.e., ACL) with respect to a single-layer strategy (i.e., SL) in terms of both bandwidth savings (for a given Γ) and wider Γ range over which QoS can be guaranteed.

Figure 4.4 plots the overall number of OFDM symbols used for PDU transmissions versus the time, when the ACL strategy ($\sigma = 2.3$ dB and $\sigma =$

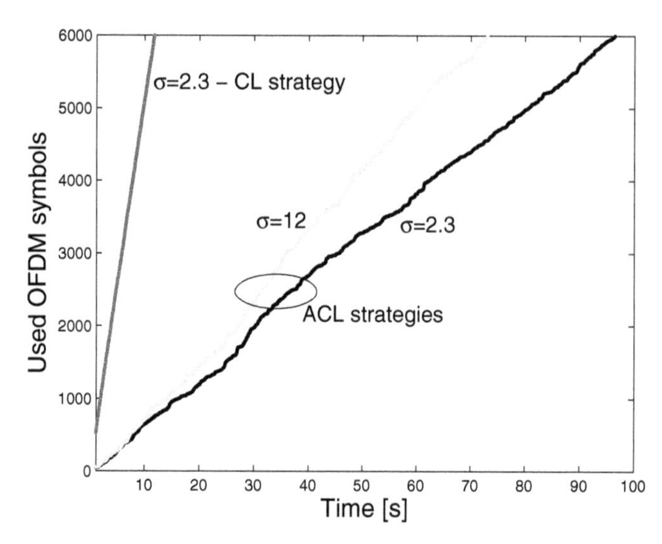

Figure 4.4 ITU vehicular A channel: overall number of OFDM symbols used versus time, for different values of σ

12 dB) and when a (non-adaptive) cross-layer (CL) strategy ($\sigma = 2.3$ dB) is used. The CL strategy optimizes both the PHY and MAC layer parameters for a mean Γ value (i.e., $\Gamma = 33$ dB) at the MT, without tracking the shadowing variations. The gap between the curves of ACL and CL strategy indicates the amount of bandwidth (in number of OFDM symbols) saved thanks to the adaptive approach. For higher values of σ (e.g., user in office), a higher amount of bandwidth is required, because there is a higher probability that the channel quality degrades and thus lower spectral efficiency modulations must be selected.

Figure 4.5 plots the expected delay, $\overline{D}$, computed from time instant at 0 s, versus time, for different requirements of $\overline{D}_{\max}$. Results show that the QoS is met even for the most stringent $\overline{D}_{\max}$ requirements and that $\overline{D}$ is not significantly affected by channel quality fluctuations. The gap between $\overline{D}$ and $\overline{D}_{\max}$ indicates that it is possible to propose other ACL strategies that trade the delay for PDU loss rate (i.e., by increasing the value of K). However, such strategies may require a larger bandwidth utilization with respect to the proposed one.

It is interesting to compare the performance of Figure 4.5 against the value of $\overline{D}$ achieved in a flat fading channel, as shown in Figure 4.6. The comparison shows that user terminal mobility cause wider fluctuations in the

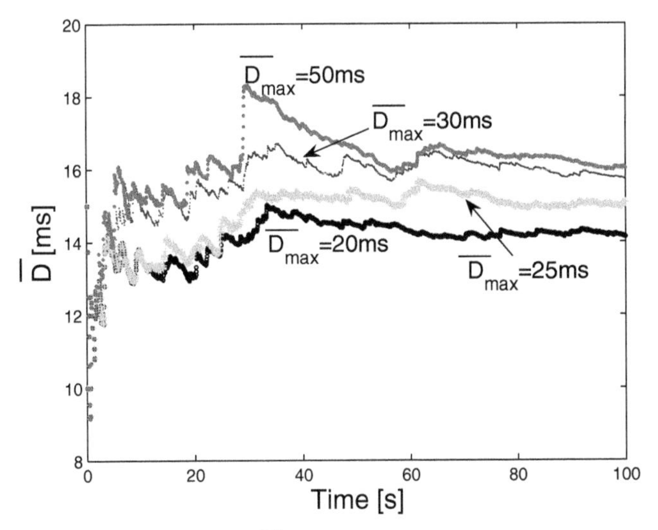

Figure 4.5 ITU vehicular A channel: $\overline{D}$ versus time, for different QoS requirements of $\overline{D}_{\max}$, $L_{\max} = 10^{-4}$

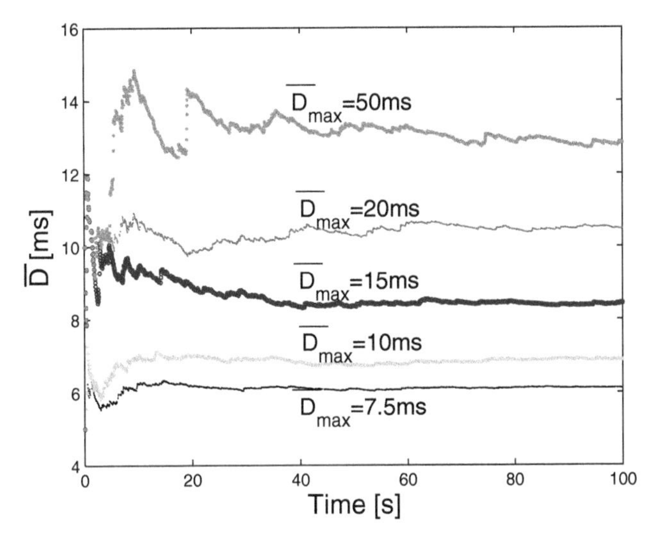

Figure 4.6 Flat fading channel: $\overline{D}$ versus time, for different QoS requirements of $\overline{D}_{\max}$, $L_{\max} = 10^{-4}$

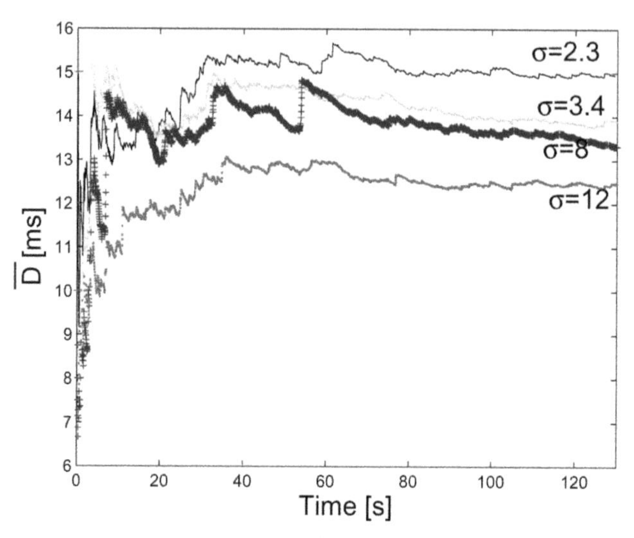

Figure 4.7 ITU vehicular A channel: $\overline{D}$ versus time, for different values of σ

MAC layer performance. Note that the performance of flat fading channel are achieved on a wider Γ range (i.e, 12–40 dB) and for a log-normal distribution centered at 26 dB, due to the very low channel frequency selectivity.

Figure 4.7 plots $\overline{D}$ versus time, for different mobility scenarios, i.e., for different values of σ. Although, the value of $\overline{D}$ tends to stabilize in an interval of few ms, it is interesting to notice that a channel with high Γ variance, due to shadowing, leads to better performance in terms of expected delay. This counter-intuitive result is due to the combination of shadowing effects and ACL strategy performance.

Figure 4.8 plots $\overline{D}$ versus time, for different service flow bandwidth requirements, i.e., for different values of λ, when $L_{\max} = 0.001$. The increase of λ leads to an increase of $\overline{D}$, making it difficult to ensure requested QoS. For values of $\lambda > 1.5$ PDU per time frame, QoS requirements cannot be guarantee anymore on the Γ range. In such case, Γ range on which QoS should be guaranteed, service flow bandwidth (e.g., to allow more than one PDU transmission per time frame) or service flow QoS (e.g., PDU loss rate and expected delay) should be renegotiated with the BS.

Simulation results show that the ACL strategy is able to promptly adapt to channel quality variations and to meet QoS requirements using the minimum amount of bandwidth.

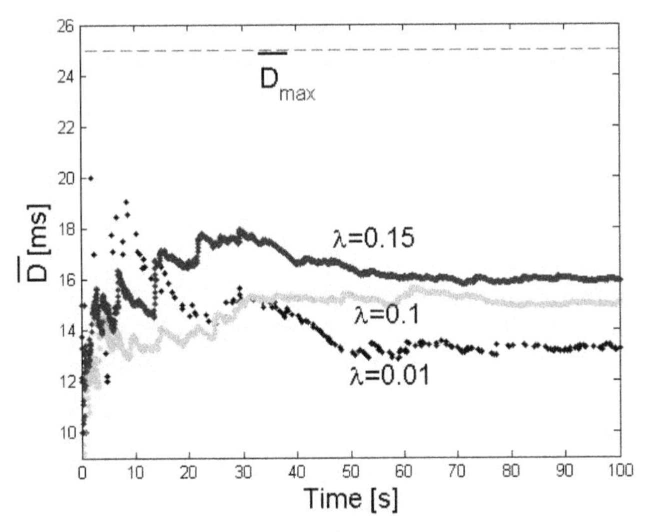

Figure 4.8 ITU vehicular A channel: $\overline{D}$ versus time, for different values of λ

4.3 DSA: Dynamic Service Flow Addition

The family of IEEE 802.16 standards permits to easily manage and guarantee the requested QoS to each connection, referred to as *service flows*, established between the BS and the MTs [18].

At the LLC sublayer of the 802.16 networks, admission control is operated via a three-way handshake procedure, referred to as DSA and service flows can be dynamically established upon request with specific QoS requirements.

The decision whether admitting or not a new service flow to the network has to guarantee that the QoS levels are maintained for the service flows already active and for the one to be added.

The DSA protocol is based on a request message, a response message, and an acknowledgement message, exchanged between BS and MT. Service flow is activated and the requested QoS is guaranteed, when the protocol terminates successfully (i.e., both the response and the acknowledgement are positive and the corresponding messages are received).

However, the unreliable nature of the radio medium may seriously compromise DSA message exchange. Thus, to ensure a successful reception of DSA messages, DSA protocol is provided with a robust retransmission strategy. When a reply to a transmitted DSA message (e.g., the request or the response message) is not received correctly within a timeout, a retransmission of the same DSA message can be attempted. Maximum number of message transmission attempts and timeout duration can be flexibility selected within a wide range of values defined by the standard. Although robust, DSA signalling can still be be blocked if all the multiple attempts of sending a request (or a response) message fail. Thus, service flow requests can be blocked due to unavailable resources to meet QoS (i.e., admission control blocking) as well as erroneous termination of DSA protocol (i.e., signalling blocking). Blocking and latency of DSA protocol are both affected by the selection of timeout values and the maximum number of attempts.

A number of works addressed the performance of data transmissions over wireless channels [112] and ARQ protocol [14] in WiMAX networks. Also, strategies to ensure QoS of activated service flows [8, 71], cross-layer approaches that adapt to the channel conditions [14, 28] and service flow scheduling [12, 66, 80, 99] have been proposed for WiMAX networks. Although relevant, these works are unable to capture the peculiar behaviour of DSA protocol and its performance, especially in a mobile environment. The problem is that a degradation of the radio channel, for instance due to mobility, could lead to an increase of DSA protocol blocking and in turn to a reduction of the overall network performance. Similar problems occur also during the message exchange for handover procedure as investigated in [128].

In this section, the performance of DSA protocol are thoroughly evaluated in a mobile networks based on OFDM physical layer in TDD. DSA is investigated in terms of signalling blocking, admission control blocking, and latency for a variety of scenarios, i.e., for different channel quality conditions, degrees of reciprocity between DL and UL radio channels, MT speeds, OFDM configurations and DSA protocol parameters.

Analytical expressions for admission and signalling blocking probabilities and for service activation latency will be derived. The probability of service activation without MT knowledge will be also derived.

A comprehensive study of the impact of the different parameters on DSA protocol performance is assessed. Simulation results aim at quantifying the impact of mobile user speeds and PHY layer parameters on MAC performance. This helps to derive useful considerations on the strategies to use to counter-react the performance degradation due to high speeds, long channel coherence time at low (pedestrian) speeds, and absence of reciprocity between DL and UL channels.

4.3.1 System Model

The physical layer considered is based on OFDM as standardized in [48]. In the OFDM physical layer, each time frame is composed by N_{ofdm} OFDM symbols. Each OFDM symbol consists of N_{fft} subcarriers. The frame duration is $T_f = N_{\text{ofdm}} \cdot T_{\text{ofdm}}$, where T_{ofdm} is the OFDM symbol duration and $T_{\text{ofdm}} = 1/\Delta F$ where ΔF is the subcarrier frequency separation. Given the maximum MT velocity and the maximum level of tolerable ICI, a minimum ΔF (i.e., a maximum duration of the OFDM symbol) is derived according to Li et al. [67] and Van Nee and Prasad [114]. The frequency separation is kept fixed to a value that usually ensures an ICI below −27 dB. Typical values of T_{ofdm} are in the order of tens of μs and time frame duration in the range of [1, 20] ms.

An OFDM symbol is considered incorrectly received if the symbol carried by any subcarrier is detected incorrectly (i.e., no FEC is assumed). The probability that an OFDM symbol is received incorrect depends on the SNR distribution over the frequency domain. The instantaneous SNR of the ith subcarrier for the jth OFDM symbol of the frame is defined as

$$\gamma_{ij} = \frac{E_S}{N_0} \|\mathbf{H}(i, j)\|^2$$

where E_s/N_0 is the average SNR and $\mathbf{H}(i, j)$ is the channel response in the frequency domain for the ith subcarrier for the jth OFDM symbol, Section 3.6.3. The distribution γ_{ij} along i depends on the channel PDP. The PDPs are chosen accordingly to the ITU channel model for pedestrian and vehicular scenarios [100]. Each path of the multipath channel is Rayleigh distributed, thus the distribution of the SNR for the subcarriers follows a Rayleigh distribution too [55]. The PDP determines the channel frequency profile; mobility and the Doppler Spectrum affect the channel response reducing the correlation of the subcarrier SNR over time, according to the speed, see Section 3.6.3.

4.3.2 DSA Signalling Analysis

Consider a wireless network based on IEEE 802.16 operating in point-to-multipoint mode, i.e., the BS communicates with a number of MTs. DL and UL transmissions are scheduled by the BS in TDD mode, within each time frame. Time frame duration is fixed and indicated as T_f.

When a DL or UL service flow is requested, admission control operations are executed. The admission control blocking is derived under the assumption that a fixed number m of service flow requests can be accommodated, i.e., each service flow requests $1/m$ of the available bandwidth in the network. Under the above assumption and assuming that the signalling latency is negligible with respect to the service flow duration, the network can be modelled as an $M/M/m/m$ queue. The probability that an arriving customer finds an $M/M/m/m$ queue full is given by the well-known Erlang B formula [38].

If the service flow is admitted, it can be activated using a three-way handshake between the BS and the MT. In the following, the BS-triggered signalling[3] is considered for establishing DL or UL service flows. For simplicity and without loss of generality, a single MT is assumed.

Upon request of a service flow, the BS performs admission control operations. If the service flow is admitted, the service flow can be activated using a three-way handshake between the BS and the MT. Signalling is triggered by the BS by sending a DSA-REQ message (i.e., dynamic service flow activation request) to the MT. Upon correct reception of the DSA-REQ message, the MT replies with a DSA-RSP (i.e., dynamic service flow activation response). Upon correct reception of the DSA-RSP message, the BS replies with a DSA-ACK (i.e., dynamic service flow activation acknowledgement). Service flow is activated at the BS after the expiration of timeout T_{10}, as described in [2, section 6.3.14.8]. This case of successful signalling is sketched in Figure 4.9(a).

The messages are sent on the primary management connection, using the most robust physical layer profile. However, due to the unreliability of the radio channel, one or more of the messages may be lost or incorrectly received. For this purpose, DSA-REQ and DSA-RSP messages may be retransmitted multiple times, upon expiration of a timeout.

Upon expiration of timeout T_7, BS can retransmit a copy of the DSA-REQ message. BS can transmit up to n_R copies of a DSA-REQ message. This case is represented in Figure 4.9(b), for $n_R = 3$.

Upon expiration of timeout T_8, the MT can retransmit a copy of the DSA-RSP message. MT can transmit up to n_S copies of a DSA-RSP message. This case is represented in Figure 4.9(c), for $n_S = 3$.

A service flow request can be blocked due to:

- admission control: BS may decide to not admit a service flow when the available bandwidth or QoS cannot be guaranteed

[3] According to the IEEE 802.16 standard, support of BS-triggered signalling is a mandatory requirement, while support of MT-triggered signalling is optional.

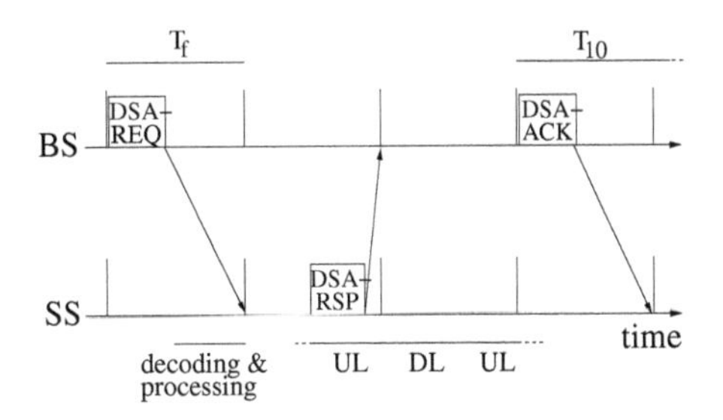

(a) Successful DSA signalling

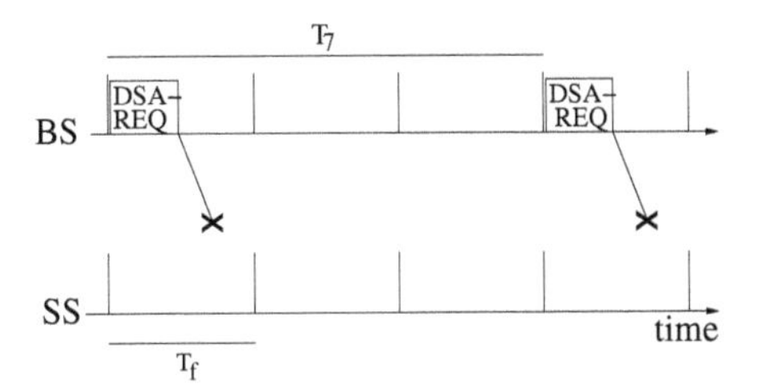

(b) Signalling blocking due to corrupted DSA-REQ messages

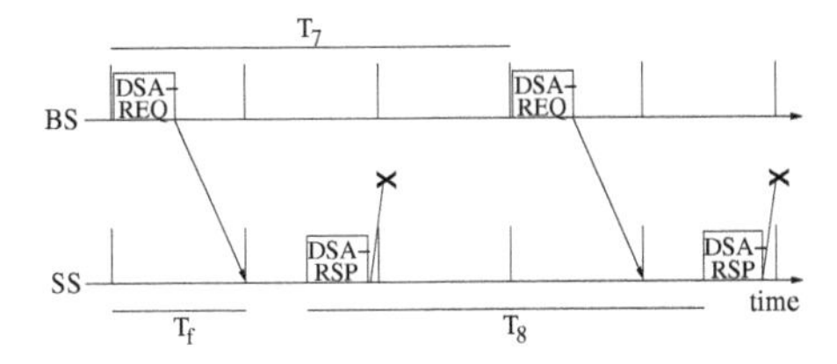

(c) Signalling blocking due to corrupted DSA-RSP messages

Figure 4.9 Time chart of DSA signalling

- erroneous termination of signalling: when the BS does not receive any DSA-RSP from the MT, the requested service flow is not activated (Figure 4.9(b))

- negative responses: for various reasons, MT (BS) can reply with a negative DSA-RSP (negative DSA-ACK). In such cases, the requested service flow is not activated.

In addition to the above cases, a service flow can be successfully activated at the BS, without MT being informed. This happens when the MT does not received any positive DSA-ACK sent from the BS (Figure 4.9(c)) and may create burdens at the MAC and upper layers.

The performance of the DSA signalling is analytically quantified in terms of blocking probability and latency, under the assumption of absence of negative responses from BS and MT: the blocking analyzed here is due to the radio transmission unreliability.

The analysis is based on the following parameters:

- p_R, p_S, p_A: expected probability that DSA-REQ, DSA-RSP or DSA-ACK is received incorrectly, respectively
- n_R, n_S: maximum number of copies of a DSA-REQ or DSA-RSP message, respectively
- T_f: time frame duration
- T_7: timeout that triggers a DSA-REQ retransmission
- T_8: timeout that triggers a DSA-RSP retransmission
- T_{10}: timeout after which a service flow is considered blocked or ready for data transmission
- λ: expected arrival rate of service flow requests
- τ: expected duration of a service flow.

DSA performance is analytically derived assuming that timeouts satisfy (see Figure 4.9):

$$T_7 \leq T_8, \quad T_{10} > (n_S - 1) \cdot T_8$$

The first assumption ensures that for a requested service flow up to n_R copies of DSA-REQ can be transmitted. In other words, two consecutive DSA-REQ for the same service flow cannot trigger a new sequence of n_S copies of the DSA-RSP message. The second assumption ensures that the BS takes the final decision about the activation or blocking of a service flow only when all the copies of DSA-RSP should have been sent.

The expected blocking due to erroneous termination of signalling, or *signalling blocking*, the expected admission control blocking and the expected probability that a service flow is activated without MT knowledge are derived. The analysis is performed assuming that the radio channel in different time frames is uncorrelated.

4.3.2.1 Signalling Blocking

Let B_R be the expected probability that no DSA-REQ is correctly received by the MT, for a service flow request, i.e.,

$$B_R = p_R^{n_R} \tag{4.8}$$

Let B_S be the expected probability that no DSA-RSP is correctly received by the BS, under the assumption that MT correctly received one or more copies of DSA-REQ for a service flow request, i.e.,

$$B_S = p_S^{n_S} \tag{4.9}$$

Thus, signalling blocking (B_{DSA}) can be evaluated as

$$B_{\text{DSA}} = B_R + (1 - B_R) \cdot B_S \tag{4.10}$$

4.3.2.2 Admission Control Blocking

Admission control blocking is derived under the assumption that a fixed number m of service flow requests can be accommodated, i.e., each service flow requests $1/m$ of the available bandwidth in the IEEE 802.16 network. Assuming that the signalling latency is negligible with respect to the service flow duration, the network can be modelled as a M/M/m/m queue. In such queue, customers are the service flows, the service time is the duration of a service flow, and the queue length is finite and equal to the maximum number of service flows that can be accommodated, i.e., m.

Admission control blocking, B_{adm}, can be computed as the probability that an arriving customer finds the M/M/m/m queue full, i.e., m customers waiting in queue. Such probability is given by the well-known Erlang B formula [38]:

$$B_{\text{adm}} = \frac{(1 - \zeta^m)}{m! \sum_{h=0}^{m} \zeta^h} \tag{4.11}$$

where ζ is the queue utilization, i.e.,

$$\zeta = \lambda \cdot \tau \cdot (1 - B_{\text{DSA}}) \tag{4.12}$$

When removing the assumption of negligible signalling latency with respect to the service flow duration, Equation (4.11) is a lower bound of the admission control blocking.

4.3.2.3 Activation without MT Knowledge

The expected probability that a service flow is activated without MT knowledge (B_{awk}) is given by the probability that MT is not able to decode correctly any of the copies of a DSA-ACK message sent by the BS, i.e.,

$$\begin{aligned}
B_{\text{awk}} &= \sum_{i=1}^{n_S} \text{prob}\,\{\text{first correct DSA-RSP is } i\text{th}\} \\
&\cdot \sum_{m=1}^{n_S-i} \text{prob}\,\{m \text{ correct DSA-RSP} \mid (n_S - i)\text{DSA-RSP sent}\} \\
&\cdot \text{prob}\,\{m+1 \text{ incorrect DSA-ACK}\} \\
&= \sum_{i=1}^{n_S} \frac{1-p_S}{1-p_S^{n_S}} p_S^{i-1} \sum_{m=0}^{n_S-i} (1-p_S)^m p_A^{m+1} p_S^{n_S-i-m} \binom{n_S-i}{m} \\
&= \frac{p_A(1-p_S)}{p_S(1-p_S^{n_S})} \left[s^{n_s} \left(\frac{1-(\frac{p_S}{s})^{n_S+1}}{1-(\frac{p_S}{s})} - 1 \right) \right]
\end{aligned}$$

where $s = (p_S + (1-p_S) \cdot p_A)$.

4.3.2.4 Signalling Latency Analysis

Signalling latency is derived under the assumption that the primary management connection has always bandwidth sufficient to transmit any DSA messages, i.e., no buffering delay of DSA messages is considered.

Signalling latency for a successfully activated service flow is defined as the time interval from the arrival of a service flow request to the activation of the service flow at the BS. Signalling latency can be decomposed into three terms: the time interval from the arrival of a service flow request to the start of the transmission of the first DSA-REQ message (L_{arr}), the time interval from the first DSA-REQ to the first correctly received DSA-RSP (L_{DSA}), and time interval from the activation of the service flow to when data transmission can take place (i.e, T_{10}), see Figure 4.9. Thus, the expected signalling latency (L) is equal to

$$L = L_{\text{arr}} + L_{\text{DSA}} + T_{10} \tag{4.13}$$

When DSA requests follow a Poisson process, the first term can be approximated[4] as

$$L_{\text{arr}} \approx \frac{T_f}{2} \tag{4.14}$$

[4] When $B_{\text{adm}} = 0$, $L_{\text{arr}} = T_f/2$.

where T_f is the time frame duration. The second term, L_{DSA}, is given by

$$L_{\mathrm{DSA}} = \sum_{i=1}^{n_R} p_R^{i-1} \frac{1-p_R}{1-p_R^{n_R}} (i-1) T_7 \left\lceil \frac{T_7}{T_f} \right\rceil + \sum_{i=1}^{n_S} p_S^{i-1} \frac{1-p_S}{1-p_S^{n_S}} (i-1) T_t + T_s \quad (4.15)$$

where T_t is the expected latency between the transmission of two DSA-RSP copies and T_s is the latency between the transmission of a DSA-REQ and the corresponding DSA-RSP, according to transmission scheduling. The first term of L_{DSA} indicates the latency from the first transmitted DSA-RSP to the first correctly received DSA-REQ. The second term is the latency required to correctly receive a DSA-RSP once a DSA-REQ has been correctly received. T_t depends on the number of DSA-RSP attempts remaining to transmit, on the timeout values for T_7 and T_8 and their ratio. An approximation is given in (4.16) in the following section.

4.3.3 DSA Analysis Validation

Blocking and latency experienced by DSA signalling are evaluated and compared under different values of n_R and n_S. Results are obtained Analytically (AN) and by Simulations (SIM). The theoretical analysis is verified with computer simulations for a block fading channel: the channel is uncorrelated among successive frames.

The considered IEEE 802.16 physical layer is as follows: $N_{\mathrm{sub}} = 1024$,[5] $\Delta f = 11.16071429$ kHz, $T_{\mathrm{ofdm}} = 102.86\ \mu$s, $N_{\mathrm{ofdm}} = 48$, $T_f = 5$ ms [127]. Signal modulation is uncoded BPSK and carrier frequency is 3.5 GHz. A single OFDM symbol is considered incorrectly received if one or more (or all) subcarriers contain an error. For such physical layer, a single OFDM symbol is sufficient to support a DSA message.

Inter-arrival times ($1/\lambda$) and duration (τ) of the service flow requests are exponentially distributed. The timeouts used for DSA are: $T_7 = 3$ ms, $T_8 = 8$ ms, $T_{10} = 20$ ms. With such selection of timeout values, it is possible to approximate T_t in (4.15) as

$$T_t = \left\lceil \frac{T_8}{T_f} \right\rceil p_R + \left\lceil \frac{T_7}{T_f} \right\rceil (1 - p_R), \quad (4.16)$$

[5] For simplicity, the number of gap and null subcarriers is 0.

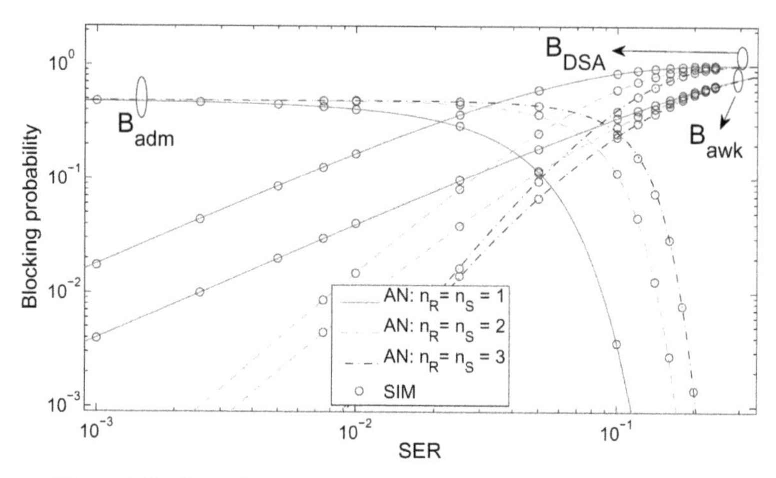

Figure 4.10 Simplified channel model: blocking probability versus SER

i.e., retransmission of a DSA-RSP message is triggered either by a DSA-REQ message when the copy of a DSA-REQ message (transmitted upon expiration of timeout T_7) is received correctly or by the expiration of timeout T_8 for DSA-RSP message. Scheduling latency in Equation (4.15) is set to $T_s = 0$ ms.

To compare the analytical results with the simulation results, an OFDM channel characterized by independent and uncorrelated and uniformly distributed OFDM symbol errors is considered. MAC layer parameters are set as follows: $n_R = n_S$, $n_R \in \{1, 2, 3\}$, $\lambda = 1$ s^{-1} and $\tau = 1$ s.

Figure 4.10 shows the different components of blocking probability versus the average OFDM SER. Analytical results closely match simulation results. This is true also for for the approximation given for the admission control blocking (B_{adm}) in (4.11).

The figure shows that signalling blocking (B_{DSA}) and admission without MT knowledge (B_{awk}) increases with SER. On the other hand, B_{adm} decreases with SER. This behaviour is due to the fact that a lower number of service flows can be activated due to the higher signalling blocking, resulting in a higher probability that resources are available for an incoming service flow request.

Increasing the number of retransmissions of control messages (i.e., n_R and n_S) helps to significantly reduce B_{DSA} and B_{awk}. On the other hand, B_{adm} increases with n_R and n_S, due to the decrease in B_{DSA}. Finally, it is important to note that B_{awk} is non-negligible, especially at high SER. This result calls

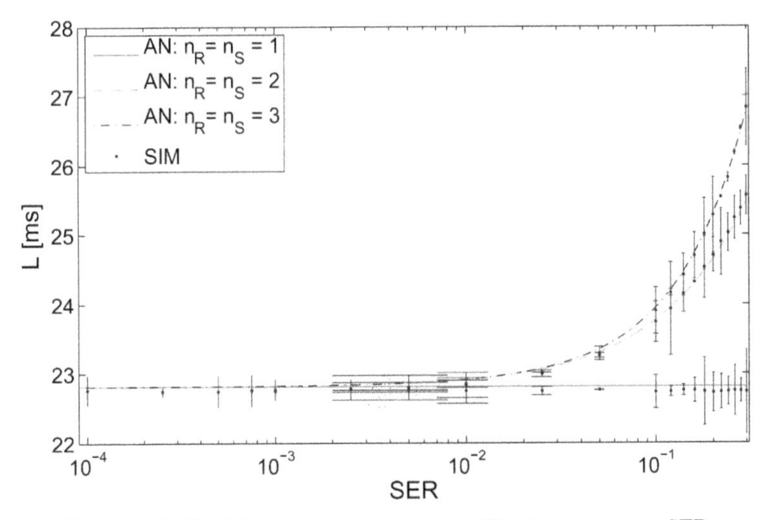

Figure 4.11 Simplified channel model: signalling latency versus SER

for protocol improvement and strategies, in order to inform the MT of the activated service flows.

The performance improvement of B_{DSA} with n_R and n_S comes at the price of an increased signalling latency (L), as shown in Figure 4.11. Analytical results well follow the simulation curves. Signalling latency is constant with SER when $n_R = n_S = 1$, i.e., only one DSA-REQ and DSA-RSP is allowed to activate a service flow. When increasing n_R and n_S, latency arises with SER, as multiple retransmissions of the same control message are allowed and may be required.

4.3.4 Performance Analysis in Mobility

In this section, DSA protocol performance is evaluated in terms of:

- Signalling blocking probability: this is the probability that no DSA-RSP is correctly received by the BS within T_{10} timeout in response to DSA-REQ message.
- Admission control blocking probability: this is the probability that a service flow request cannot be admitted due to lack of available resources.
- Signalling latency: this is the time interval from the arrival of a service flow request to the instant in which data transmission has taken place.

A simplified admission control mechanism is derived under the assumption that a fixed number m of service flow requests can be accommodated, i.e., each service flow requests $1/m$ of the available bandwidth in the IEEE 802.16 network.

To quantify the performance, MAC and PHY layer are implemented as described in Section 4.3. in a IEEE802.16 C/C++ custom-made event-driven simulator. Simulation results are collected using the following configuration, unless otherwise indicated. Parameters of the OFDM physical layer are: $N_{\text{fft}} = 1024$, $\Delta F = 11.16071429$ kHz, $T_{\text{ofdm}} = 102.86\ \mu\text{s}$, $N_{\text{OFDM}} = 48$, $T_f = 5$ ms [127]. With this selection of N_{fft}, each DSA message can be accommodated on a single OFDM symbol. Signal modulation is uncoded BPSK and carrier frequency is 3.5 GHz.

At the MAC layer, the timeouts are set to: $T_7 = 3 \cdot T_f$, $T_8 = 3 \cdot T_f$, $T_{10} = 4 \cdot T_f$. Scheduling latency for transmitting DSA messages (i.e., buffering delay) is considered negligible. Latency due to signal and information processing at the PHY layer is accounted as indicated in Figure 4.9, i.e., half time frame is required for decoding the message and generating a reply. Maximum number of copies of DSA-REQ and DSA-RSP messages are $n = n_R = n_S \in \{1, 3\}$. Inter-arrival times $(1/\lambda)$ and duration of the service flow requests are exponentially distributed. Inter-arrival rate of DSA requests is set to $\lambda = 20\ \text{s}^{-1}$ and expected duration of service flows is set to 1 s. Each service flow requests 20% of the available bandwidth in the IEEE 802.16 network. Effects of MT mobility are taken into account by implementing a channel model with Jakes Doppler spectrum. The ground speed v of MT with respect to the BS is selected in the range $[5, 300]$ km/h. For comparison purposes, a single-path block-fading channel without mobility is also considered and indicated as $v = 0$. DL and UL channels are assumed to be independent.

4.3.4.1 Impact of Mobility

Figures 4.12–4.14 show the impact of MT mobility in terms of signalling blocking, admission control blocking and latency, respectively, when $n = 1, 2$, and 3. Mobility deteriorates the channel performance and degrades the DSA performance in terms of signalling blocking and latency. However, when MT speed exceed 50 km/h (i.e., from 50 to 300 km/h), DSA performance do not degrade any further. Signalling blocking can be improved by resorting to higher number of transmission attempts (n), to detriment of an increased latency. Admission control blocking is reduced with mobility. This happens because the number of service flow requests blocked by erroneous

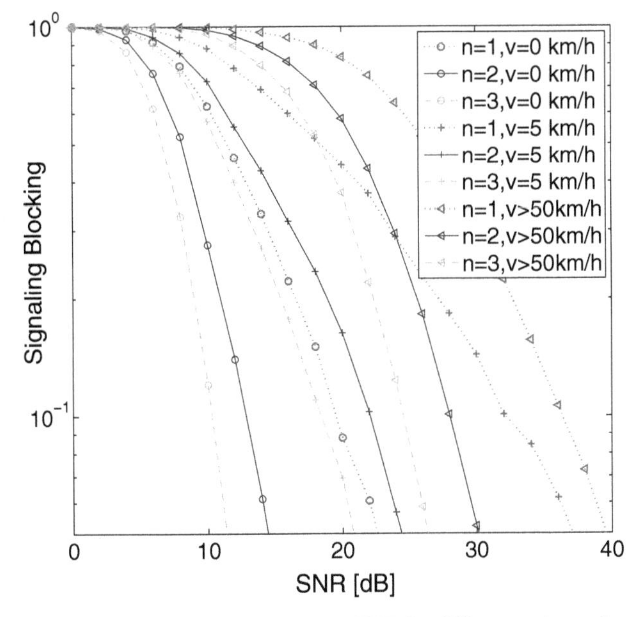

Figure 4.12 Signalling blocking versus SNR for different values of v and n

termination of signalling increases with MT speed and, thus, the network load decreases lowering the probability of finding a full admission queue.

4.3.4.2 Impact of Correlation between UL and DL Channels

Figures 4.15 and 4.16 compare the signalling blocking and the latency, respectively, experienced at pedestrian speed ($v = 5$ km/h), when the DL and UL channels are reciprocal (i.e., DL and UL are considered as a single fading channel) or independent fading processes. In Figure 4.15, independent block-fading channels (with no time correlation) are also included for comparison. The results indicate that the reciprocity is beneficial when $n = 1$ (no retransmissions) for any value of SNR. When $n > 1$, reciprocity benefits are limited to low SNRs and may compensate the negative effects of channel correlation which has worse performance of the case of block-fading channels (no time correlation).

Figure 4.16 shows that the correlation between the DL and UL channels allows to greatly lower the latency of the admitted service flows, especially at low SNR.

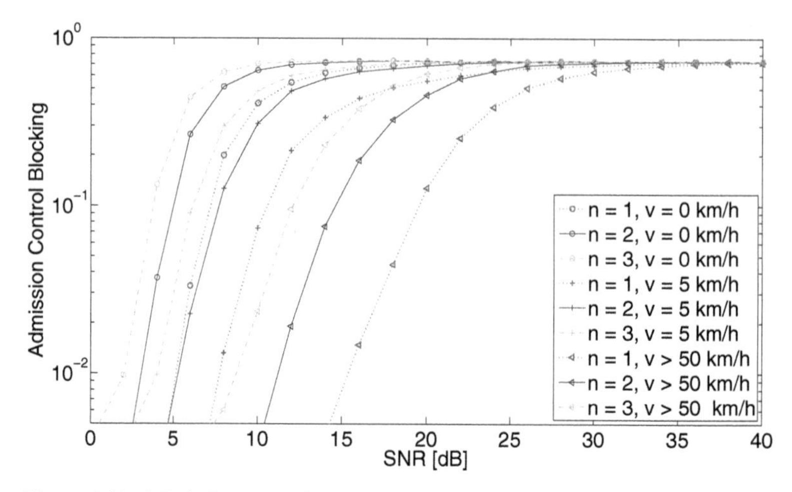

Figure 4.13 Admission control blocking versus SNR for different values of v and n

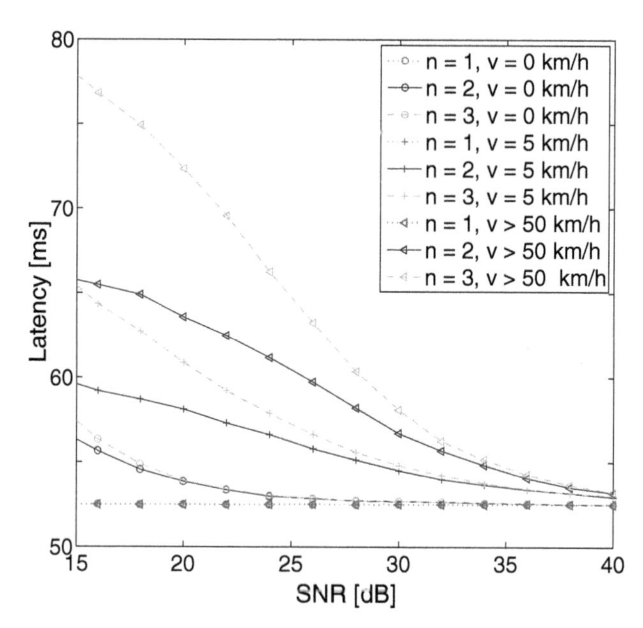

Figure 4.14 Latency versus SNR for different values of v and n

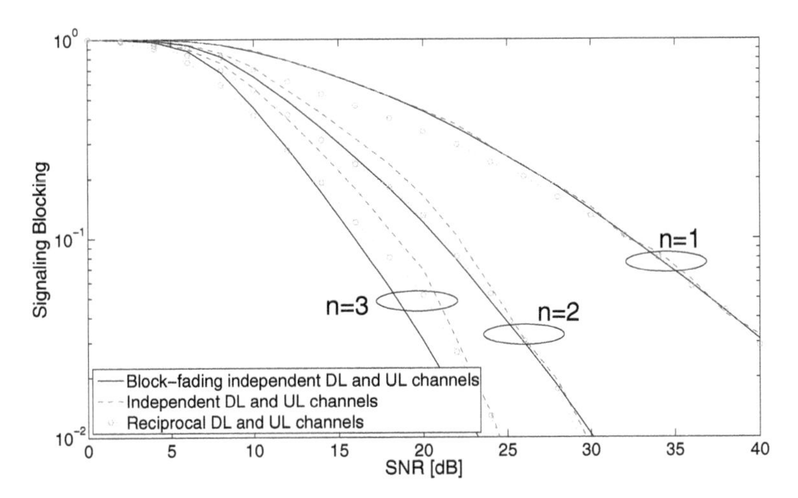

Figure 4.15 Signalling blocking versus SNR for v = 5 km/h, when UL and DL channels are either independents or fully reciprocal

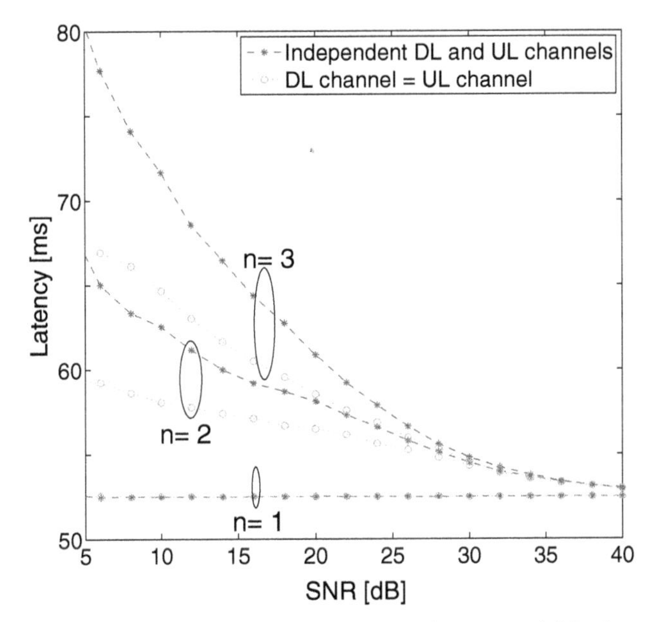

Figure 4.16 Latency versus SNR for v = 5 km/h, when UL and DL channels are either independents or fully reciprocal

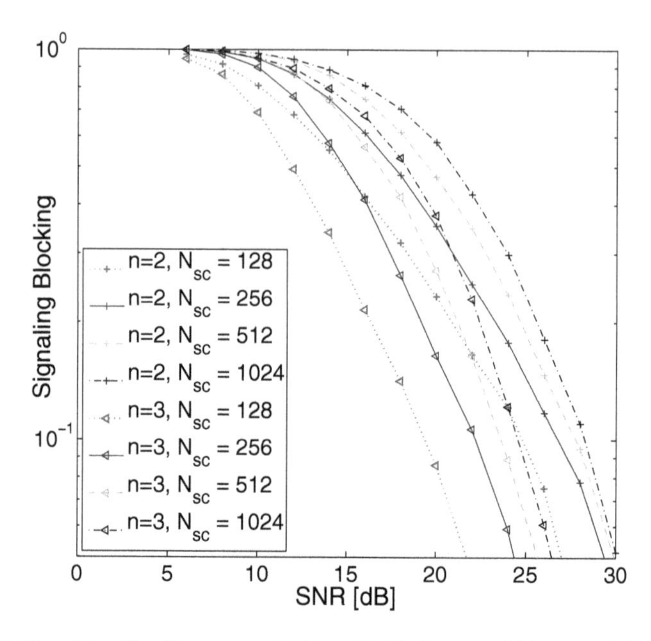

Figure 4.17 Signalling blocking versus SNR = 20 dB, for $v = 50$ km/h and different values of n and N_{fft}

In the presence of time correlation, the reciprocity of DL-UL channels is beneficial. This is due to a lower coherence time of the error process when DSA messages are sent over a single fading channel: the coherence time of the independent channels is lower bounded by the coherence time of the reciprocal DL/UL channels. Loss of reciprocity between DL and UL channels can be recovered with calibration techniques [40] which could be available in future MTs.

4.3.4.3 Impact of the Number of Subcarriers per OFDM Symbol

In IEEE802.16 scalable OFDM [127], ΔF is fixed and different bandwidths can be exploited varying the number of subcarriers (N_{fft}), thus, the impact of N_{fft} over DSA is analyzed. For $N_{\mathrm{fft}} < 1024$, the DSA messages are spread over two (or more) consecutive OFDM symbols. A DSA message is received incorrectly when at least one of OFDM symbols is in error. In Figures 4.17 and 4.18, the signalling blocking is evaluated as a function of SNR for various value of N_{fft}, when $v = 50$ km/h and $v = 300$ km/h, respectively. At 50 km/h, the correlation of the channel is favourable: if an OFDM symbol is received

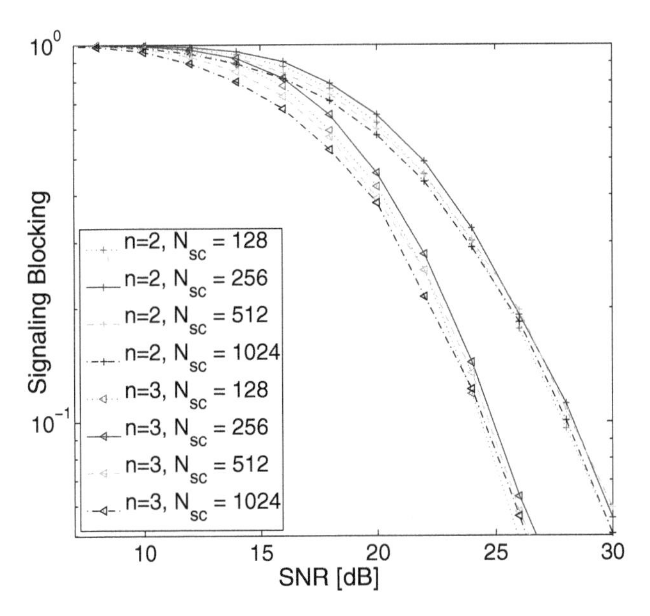

Figure 4.18 Signalling blocking versus SNR = 20 dB, for $v = 300$ km/h and different values of n and N_{fft}

correctly, with high probability the consecutive OFDM symbols are received correctly too, leading to a correct reception of a DSA message. Thus, the channel correlation, combined with the small values of N_{fft}, lowers the signalling blocking. In Figure 4.18, at 300 km/h, the difference among the curves at different N_{fft} is greatly narrowed because, at high velocities, the channel varies sensibly during consecutive OFDM symbols and the probability that two or more consecutive OFDM symbols are received incorrect is lower.

4.3.4.4 Impact of Time Frame Duration and DSA Timeouts

In scalable OFDM, the frame duration, T_f, is also configurable; in particular, short T_f with low N_{fft} have high overheads and should be avoided [127]. Figure 4.19 shows the impact of T_f on the signalling blocking probability as a function of the SNR, for $v = 5$ km/h. The signal blocking probability decreases not only with an higher number of retransmissions but also with longer frame duration. Indeed, by using long time frames, the channel correlation can be reduced, with a beneficial effect on the signalling blocking. The positive effect of T_f duration is particularly evident at high SNR – where the

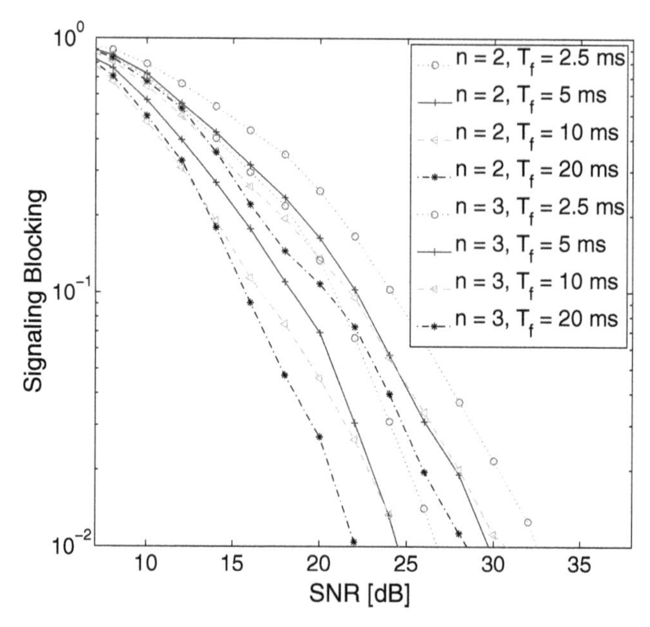

Figure 4.19 Signalling blocking versus SNR for $v = 5$ km/h and different values T_f

channel performance is dominated by the fading coherence time with respect to the additive white noise.

Figure 4.20 shows the latency normalized to number of time frames, for $v = 5$ km/h and various value of T_f. Correlation experienced at low T_f plays a positive effect on the latency, i.e., if a message is received correctly so do the following ones with high probability. Thus, low values of T_f are preferable for a fast activation of service flows, However, in a practical case, small values of T_f may lead to a loss of throughput due to overheads (e.g., for synchronization and data maps). Since the normalized latency for various T_f is comparable while T_f may up to eight times longer, it is evident that low values of T_f are preferable for a fast activation of service flows. However, in a practical case, small values of T_f may lead to a loss of throughput due to overheads (e.g., for synchronization and data maps).

Vice versa, the use of long time frames allows to exploit the time diversity and, thus, increase the probability of a successful reception to detriment of a higher latency.

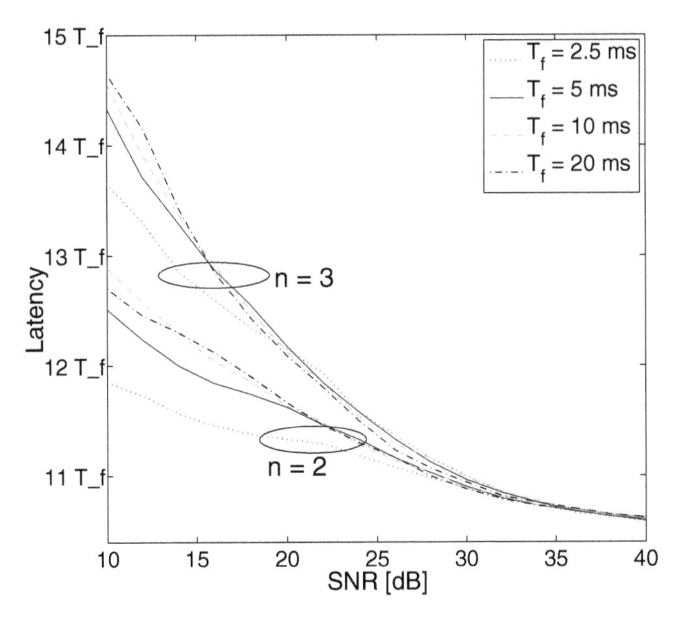

Figure 4.20 Latency versus SNR for $v = 5$ km/h and different values T_f

4.3.4.5 Impact of Timeout Duration

In addition to time frame duration, the channel correlation experienced by successive transmission of DSA messages can be controlled by varying the duration of DSA timeouts. Figures 4.21 and 4.22 show the impact of timeout duration on DSA performance, for increasing values of SNR and $v = 5$ km/h. Consistently with the results in Figure 4.19, the lower signalling blocking is achieved for longer timeout duration (i.e., $T_7 = T_8 = 4$ time frames in the figures) as time diversity can be exploited. Even in this case, the price to pay for the improved blocking is an increase of the latency.

4.3.4.6 Impact of Network Load

Figure 4.23 shows the admission control blocking experienced by a MT moving at $v = 5$ and 300 km/h, respectively, and with a channel SNR of 20 dB, for increasing inter-arrival rate λ. The admission control blocking increases with λ, but also with n. The reason is that, for larger values of n, a higher number of service flows are admitted, leading to an increase of the effective network load. The increase of admission control blocking with n is particularly evident at MT speed of $v = 300$ km/h.

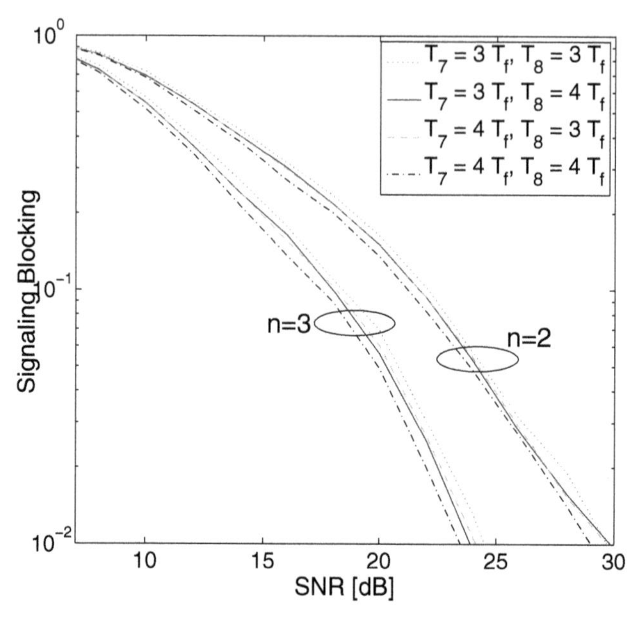

Figure 4.21 Signalling blocking versus SNR for different values of timeouts and $v = 5$ km/h

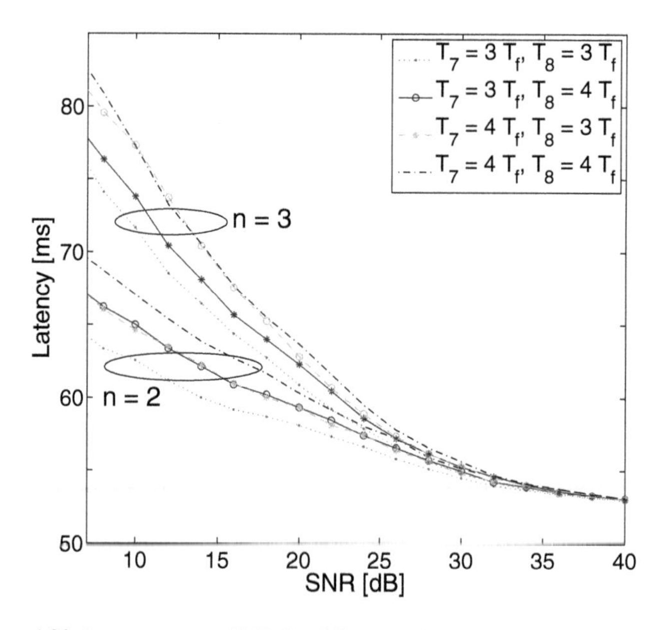

Figure 4.22 Latency versus SNR for different values of timeouts and $v = 5$ km/h

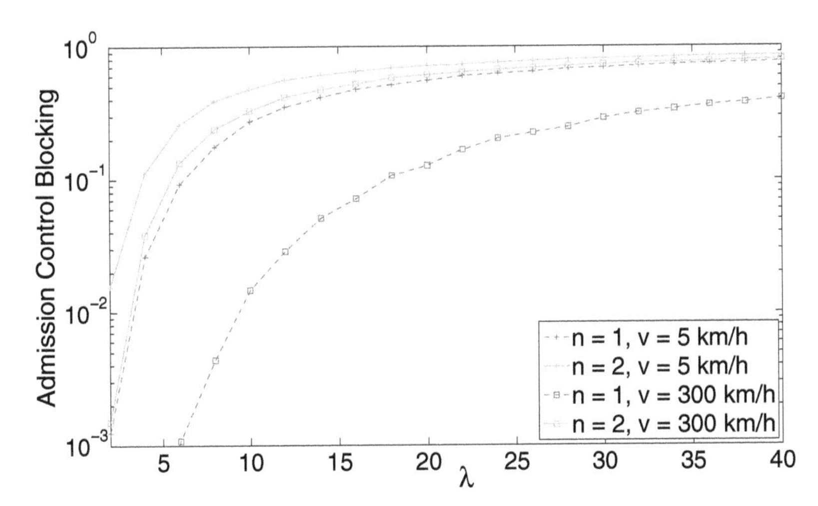

Figure 4.23 Admission control blocking versus λ for SNR = 20 dB and v = 5 and 300 km/h

Results indicate that robustness of DSA protocol can be achieved by increasing the number of message transmission attempts, at the expenses of an increased latency. This is especially important for fast moving MT. Indeed, mobility has been shown to have a detrimental effect on the DSA performance. Contrary to fast moving MT, slowly moving MT (e.g., pedestrian speed) suffers from the effect of the long channel coherence times and from the independence of DL and UL channels. To compensate the negative impact of the long channel coherence time on the signalling blocking, time diversity could be exploited, equivalently attainable by selecting longer time frames or longer duration of timeouts, at the expenses of the latency. In addition, a careful selection of the number of OFDM subcarriers (e.g., as in scalable OFDM), that are used to carry the DSA message, may help to reduce signalling blocking and latency. Finally, these results help to properly configure DSA parameters (i.e., timeouts and number of retransmissions) to achieve a requested level of DSA blocking and latency.

4.4 Conclusions

This chapter has been focused on the analysis of a IEEE802.16e MAC.[6] In particular, firstly an ARQ Adaptive Cross Layer strategy that jointly optimizes PHY and MAC parameters to meet users' QoS was presented. ARQ ACL strategy is based on a queueing theory model. The theoretical results were shown to perfectly match the simulation results. ACL strategy was tested in different scenarios of user mobility.

Secondly, the performance of the Dynamic Service Addition (DSA) protocol has been analyzed. An extensive performance evaluation of the DSA protocol has been carried out through theoretical analysis. Theoretical results have been verified with computer simulations for block fading channel and they perfectly matched the simulations. Performance evaluation has been performed also for a variety of scenarios. Signalling blocking, admission control blocking, and latency experienced by DSA protocol have been quantified for different channel and mobility conditions and various PHY and MAC parameters. DSA protocol parameters thus can be trimmed to meet QoS requirements especially for highly moving MTs to react to detrimental signal shadowing.

The ARQ protocol will be used with dynamic MIMO schemes in Chapter 5 and the MIMO signal processing gain will be jointly exploited with the protocol gain coming from the ARQ process.

The integration of DSA and ARQ with MIMO schemes will be discussed in Chapter 6 where the impact of introducing MIMO protocol gain in current network architectures will be studied.

---

[6] These conclusions can also refer to a generic wirelees network with an OFDM PHY layer and a connection-oriented MAC layer, such as LTE-A.

Part II

Cross-Layer MIMO

5

MIMO Protocol Gain

5.1 Introduction

During the last decade, the research community has witnessed the transition from MIMO as a theoretical concept to a practical system component for enhancing the performance of wireless networks. Point-to-point MIMO over a single radio link promised large gains for both channel capacity and reliability, essentially via the use of spatial multiplexing and space diversity, respectively; the extra spatial degrees of freedom coming from MIMO signaling were and are still exploited for expanding the dimensions available for signal processing and detection, acting mainly as a pure PHY-layer performance booster.

However, several difficulties have been encountered when very high SEs are attempted. This is mainly due to channel propagation characteristics (loss of MIMO channel ranks) and high correlation of the antenna elements mainly at the MT side, where the small dimensions pose serious challenges in obtaining independent fading statistics on each antenna. Further, in single-user MIMO, the link layer protocol cuts the performance benefits of MIMO antenna because it does not have full awareness of the MIMO capabilities, even if allowing for greater per-user data rate or more reliable communications.

The recent development of CL techniques, aiming at the joint optimization of the PHY and MAC layers, has shown that noticeable performance improvement is at hand. The research community is focusing on multi-user designs to attain a protocol gain, evolving from the classic single-link MIMO techniques.

An important benefit of this approach comes from the opportunistic MU scheduling. As opposed to point-to-point communication, multi-user communications introduces several new features, like cooperation, superposition of information, use of side information, joint source and channel coding, interference management, and competition techniques [10, 63, 75, 111].

Fundamental recent results in this area have indicated that PHY layer (modulation/coding), link layer, resource allocation and scheduling are deeply connected. A remarkable improvement can be achieved when all these parameters are considered with the objectives of capacity maximization and best resource exploitation.

5.2 MU MIMO State of the Art

Multiple antenna communications play an important role for achieving the high data rates required by the increasing traffic demand to be supported in future wireless networks.

When, from a processing gain exploited on a single link, MIMO is used jointly with higher layers protocols, the SU MIMO limitations discussed in Chapter 2 can be mitigated and information theory has shown that consistent gains can be achieved when MIMO is applied in MU scenarios [27, 43, 87, 93, 120, 132].

MU MIMO can achieve capacity gains not only via multi-user multiplexing schemes but also transmitting with multiple antennas at the BS towards different users simultaneously (MU-MISO systems) and with transmission from multiple MTs (even with a single antenna) towards the BS (MU-SIMO systems). IEEE802.16 Uplink Collaborative scheme falls inside the latter of the above-mentioned classes.

The scheduling of the set of the best users in the cell maximize the sum-rate, exploiting the *multi-user diversity*. Hybrid SM transmission can be used to counterfight the problem of low MIMO channel rank for a certain MT jointly scheduling different users: the equivalent MIMO matrix rank can be preserved due to the statistical independence of the channels of widely separated MTs.

However, from an information theoretic point of view, the sum-rate capacity in MU MIMO DL Channel is achieved with Dirty Paper Coding (DPC) [19], which is an interference cancellation technique combined with user scheduling and power control. The complexity of signal processing and the heavy feedback needed for DPC are too high to be implemented in a real system and a pletora of solutions have been proposed to approach, as close as possible, the capacity upper-bound represented by DPC.

Another limitation comes from the CSI feedback. When CSI feedback is feasible, its overhead can affect the performance of the network in terms of throughput. The CSI feedback is challenging for two reasons: feedback channel capacity is proportional to the number of users and antennas and, for mobile MTs, the information quickly becomes outdated. Partial (or no) feedback schemes have been proposed, see [21,54,101,125], just to cite some recent works.

User selection mechanisms are also required. This calls for a problem solver with a search space proportional to the number of users currently admitted in the network. Brute force search can be not feasible and trade-off solutions have been proposed, as in [123].

A simple MU MIMO system design with low feedback was proposed in [116]: Opportunistic Random BeamForming (ORBF). ORBF needs an high number of users in order to have an effective system throughput gain. In [57], the issue of a low/realistic number of MTs has been addressed using multiple random coefficients inside a single time slot but exacerbating the problem of the feedback rate. In [59], a memory-based approach is proposed to exploit the time correlation of the channels when the number of users in the cell is low.

The problem of feedback can be totally avoided resorting to open loop MIMO schemes, such as STBC or SM, which do not require CSIT. These schemes can be exploited to achieve diversity and/or rate gain. The optimal scheme can be selected dynamically, based on the channel status and on the spatial characteristics of the MIMO channel. A switching system based on the Demmel condition number of the channel matrix was presented in [41] and discussed in Section 3.5.1.

Equal SE can be optimally chosen based on constellation distances expected at the receiver. The decision is taken at the receiver based on locally estimated CSI and then sent back to the BS. Adaptation to the channel state can be challenging in very fast channels since the Demmel condition number can be subjected to quick variations, as in the simulation reported in Figure 3.7. When variable SE MIMO schemes are available, the selection can be based on a minimum BER to be guaranteed at the PHY layer, e.g. as in [74] and in Section 3.5.

This chapter presents the analysis of a system with ARQ and MIMO schemes jointly integrated and optimized over a narrowband Rayleigh channel. Several retransmission strategies are proposed for MU networks, as detailed in Section 5.4.

The reference MIMO system is chosen as a 2×2 system as it constitutes the baseline for the next generation wireless networks. When more antennas are available, more than two users can be served balancing the trade-off between SM and STBC, e.g. using a layered STBC [4,41,49], see Section 3.3.1.

Retransmission strategies can be designed for larger antenna sets or the antenna elements can be partitioned in sets of two antennas and MIMO-ARQ protocol applied independently on each set.

The work proposed in this chapter stems from the results obtained in [22, 23] where an optimal ARQ retransmission strategy has been defined for a single user using SM and STBC MIMO schemes.

In MU scenarios, the MIMO-ARQ transmission protocol addresses the previous drawbacks of MU MIMO exploiting multi-user diversity via random user selection without any need of CSI feedback. Open-loop MIMO schemes are used – with the added hybrid SM scheme – jointly with Multi-User ARQ and the strategy designed in [23] has been extended for an MU scenario.

The SM-STBC switching allows for preserving diversity order in the low SNR region in a smoother way than the strategies presented in [41, 74]. The system proposed allows for sending data streams to different users in SM: hybrid SM. When transmitting to two different users, the users to be coupled together can be chosen randomly or being informed by the feedback of the post-processing SNR from the MTs to the BS. Hybrid SM transmissions gives the opportunity to send a packet to a new user while retransmitting the wrongly decoded packet to a previous user: the users are served with a lower average delay and the total system throughput is leveraged.

Multi-User ARQ has been investigated for optimizing multicast/unicast flows in broadcast transmissions [64, 65].

An optimal strategy for single and multi-user cases is defined. The packet scheduling is executed in RR fashion. The system serves a pool of users in DL. The users are served with an implicit RR scheduler with TDMA; fairness among users is thus guaranteed (as in [116]). However, the system design easily permits to introduce QoS aware scheduling, taking advantage of the increasing capacity achieved via the MIMO-ARQ transmissions, as visible in Figure 5.1.

The MIMO-ARQ protocol decides the packet scheduling after the BS has decided the AMC profile based on the long-term SNR of each MT. Here, the analysis is conducted for homogeneous AMC profiles among all the MTs.

In Section 5.3, the work from [23] is presented, as it is the basis for the further developments in MU scenario. Then, in Section 5.4, the optimal antenna allocation and MU retransmission strategies are analyzed. In this case, it is possible to enhance the performance of the system with an optimized receiver which stores the unintended packets (the packets that were sent to other users in the previous transmission slots). This receiver is later referred to as: Packet Overhearing (PO) and detailed in Section 5.6. In Chapter 6, the architectural issues related to the introduction of the MIMO-ARQ protocol proposed in this chapter will be discussed.

5.3 ARQ Strategies for Single-User MIMO

This section presents the work of Carvalho and Popovski [23]. The design proposed herein is then further generalized for the Multiple User case in [78].

The channels between the transmitting and receiving antennas are modelled with flat fading Rayleigh statistics and are considered constant over the retransmission period. The packets are composed by a number of data symbols plus an appended Cyclic Redundancy Check (CRC) to enable error detection at the receiver. The data is 4-QAM modulated and the duration of the transmission slot is fixed. During each slot two packets are sent, one from each transmitting antenna. The MIMO system is 2×2. On each antenna, the packet is coded/modulated independently and it can be independently retransmitted. The retransmission strategy has the following characteristics:

- retransmitted packets have the same symbol content of the original message
- sign and conjugation operations are allowed.

The retransmission strategy is decided at the MT; one additional bit of feedback is required to indicate if the antenna allocation is to be switched, together with another bit (for each packet) indicating if the packet was received or not (ACK/NACK). The feedback is received error-free and without delay. The CSI at the receiver is assumed perfectly known.

The SU strategy is summarized as follows:

- when at time t only one packet is in error, the erroneous packet is retransmitted at $t + 1$ from one of the antenna and a new packet is transmitted from the other antenna; antenna switching is signalled back from MT
- when both packets are erroneously decoded, both are retransmitted using Alamouti coding [5]
- when both packets are decoded correctly, in the next time slot two new packets are sent.

The receiver is a SIC-MMSE receiver with Packet Combining (PC): the receiver stores the signals containing the packets decoded with errors until the current time slot and removes the contribution of the correctly decoded packets. Those past signals ensure a higher probability for correct decoding of the transmitted packets.

The work in [23] assumes the same AMC level. The aim of the design is to maximize the throughput. Two designs are proposed: one design based on post-processing SNRs at the output of the SIC-MMSE receivers at the MTs and one based on an approximation of the throughput.

5.3.1 Single-User (SU) Protocol Operation

The notation used in the following presentation is reported:

- X_A is a packet (data plus CRC symbols), in the first time slot, X_A is transmitted from antenna one and X_B from antenna two
- x_a is the single modulation symbol inside packet X_A
- $\mathbf{H}(t)$ is the 2×2 MIMO channel matrix
- $h_{ij} \in \mathbb{C}$ is the channel coefficient from transmit antenna j to receiver antenna i
- $\mathbf{H}_A$ or $\mathbf{H}_B$ are the column vectors defined respectively as the channel coefficients from the transmitting antenna for packet X_A or X_B to the two antennas at the receiver.

The BS always optimistically starts with SM transmission. If two packets are decoded, the receiver sends two ACKs and the system remains in SM mode.

Assume that X_A and X_B have been transmitted in the first time slot. The first case applies when X_A and X_B are decoded with errors. At time $t+1$, Alamouti coding starts, antenna allocation is switched – X_B is retransmitted from antenna 1 and X_A from antenna 2 – and sign and conjugation operations are applied, as in [5]. The system design permits to seamless switch between SM and STBC modes. This appears more flexible than the solution proposed in [41] or in [74].

The second case applies when only one packet, say X_A, is decoded with errors: X_A is retransmitted from the weakest antenna because if it were re-transmitted from the strongest antenna, the total information available for decoding the packet after two transmissions could have been unnecessarily large.

The first two schemes are providing gain if the channel is highly correlated between two successive time slots (see Section 5.7.4 for an analysis in fast-varying channels for SU and MU networks). The MIMO channel $\mathbf{H}(t)$ is assumed constant over two transmission slots. If a more complex MIMO system is available, the retransmission format could be applied at each couple of antennas independently; it could happen that two antenna are swapped while the other transmit other packets in SM mode or LSTBC schemes could be applied as in Section 3.6.

Those schemes are optimal (according to the design criterion) when X_A and X_B are new transmitted packets at time t. When either X_A or X_B are (re)transmitted before time t, the schemes are not always optimal. The design aims at maximizing the system throughput. In the case of retransmissions, the throughput cannot be expressed easily in closed-form. Two retransmission designs are proposed with using an approximation of the actual throughput: one based on throughput and one based on post-processing SNR at the output of SIC-MMSE receiver.

5.3.1.1 Throughput Based Design

Since $\mathbf{H}^P(t+1) = \mathbf{H}(t)$ (correlated channel over two time slots assumption), the receiver calculates the predicted SNRs for both streams at time $t+1$: $\gamma_K^P(t+1)$, where K stands for the particular stream and $K \in A, B$. The predicted throughput at time $t+1$ can be computed as

$$\mathrm{Thr}^P(t+1) = \mathrm{Thr}_A^P(t+1) + \mathrm{Thr}_B^P(t+1) \tag{5.1}$$

where

$$\mathrm{Thr}_K^P(t+1) = r_K(t+1)[1 - \mathrm{PER}_K(\gamma_K^P(t+1)]$$

and $r_K(t+1)$ and $\mathrm{PER}_K(t+1)$ are the nominal bit rate and Packet Error Rate (PER) for stream K at $t+1$. The MT decides to swap antenna allocation in order to obtain the maximum throughput evaluating (5.1) for the two allocations possible. The reliability of the decision decreases as the channel correlation decrease. Equation (5.1) do not express the real throughput of the system since it does not take into account the whole retransmission. This criterion is also valid when both streams have different AMC levels.

When the two streams have the same modulation and coding profile, maximizing (5.1) is equivalent to minimizing the sum of the PER_A and PER_B. When closed-form PER expressions are not available for the PHY profile in use, look-up tables can be used.

5.3.1.2 SNR Based Design

In the region where $\mathrm{Thr}^P(\gamma)$ is a concave function of γ, maximizing (5.1) is equivalent to maximizing (5.2) and antenna switching is decided for maximization of the sum of the SNRs of both streams at the SIC-MMSE receiver output:

$$\gamma^P(t+1) = \gamma_A^P(t+1) + \gamma_B^P(t+1) \tag{5.2}$$

Maximizing (5.2) is equivalent to minimizing the difference, over the retransmission slots, of the sums of the channel coefficients from the transmitting antenna to both the receiver antennas:

$$\min\left(|\chi_A^P(t+1) + \chi_B^P(t+1)|\right) \tag{5.3}$$

where

$$\chi_X^P(t+1) = \sum_{s\geq 0}^{L} \|\mathbf{H}_X(t-s)\|^2 + \|\mathbf{H}_X(t+1)\|^2 \tag{5.4}$$

and L is the number of retransmissions elapsed. The previous relations mean that if packet X_A is decoded with errors, retransmission is done from the weakest antenna. When packets X_A and X_B are decoded with errors, antenna allocation is switched or not, depending on the relative value of $\chi_A^P(t+1)$ and $\chi_B^P(t+1)$.

After antenna allocation has been decided, the retransmission has to decide if sign or conjugate operators are to be used. Contrary to the case where both packets are decoded with errors and are retransmitted following the structure of an STBC using sign and conjugation operators, if only 1 packet is in error, the packet is retransmitted unchanged. In any case, the format is unique and is known a-priori by the transmitter: the transmitter only needs to know the antenna allocation.

5.4 MU MIMO-ARQ

The proposed MIMO-ARQ CL protocol without feedback from the MTs to the BS integrates the characteristics of SM and STBC MIMO schemes and ARQ in a MU networks.

The system proposed in [23] could be straightforwardly applied in MU systems in a TDM fashion. For the current user scheduled, the system retransmits the packets until they are received or until the maximum number of retransmissions are achieved, then the next user in the scheduler queue is served with the same strategy. However, a MT in bad channel state forces the system to wait until the ARQ protocol elapses the full retransmission time, which is not efficient.

The MU MIMO-ARQ protocol integrates the ARQ process and open-loop MIMO schemes in an MU system with an implicit RR scheduler with the ability to send packets in hybrid SM, so that other users can be served while the wrong packet directed to a user in bad channel condition is being retransmitted. The system design allows for integration of any kind of scheduler considering QoS attributes of the traffic flows.

Several retransmission strategies are proposed. Firstly, two strategies are studied. Assume that in the first time slot two packets are transmitted in SM. A first strategy is to retransmit the wrong packet with double power from the best antenna (strategy S1) while the other antenna is switched off. The other strategy is to retransmit the wrong packet from both antenna to the same user in spatial repetition coding (strategy S2).

A more comprehensive strategy is proposed in order to exploit the MU diversity (strategy MU) which observes the following: when there is one packet in error for user (1), the packet is assumed to have been transmitted from the weaker antenna. The antenna allocation is thus deterministically switched and the packet retransmitted from the (assumed) stronger antenna. A packet for a new user $(u + 1)$ is transmitted on the other antenna. The antenna allocation follows the opposite principle respect to [23] and aims to give the maximum SNR of the recombined signal in order to get rid of the bad user as quickly as possible. This strategy permits to have the lowest service time for the user (1), reducing the average delay in the system.

The hybrid SM transmission (data streams sent to different users) attains higher throughput exploiting the multi-user diversity: indeed, while the wrong packet is being retransmitted, other users have the chance to receive a packet. The new user is chosen randomly. The performance of the strategy is compared to the case where the best user is chosen in order to maximize the throughput (i.e. the user with the highest post-processing SNR at the output of the receiver for the antenna to be used).

During hybrid SM, ARQ is used with two different users; this constitutes a Multi-User ARQ. Multi-User ARQ was previously introduced at MAC layer for multiple unicast flows in [64, 65] where each MT stores not only the packets directed to the local MT but also the packets directed to the other users (unintended packets). ACKs are sent back by each user for both types of packets (intended and unintended). The aggregate throughput efficiency is enhanced via wise selection of the users and joint encoding of retransmitted information. A joint network and channel coding with Multi-User ARQ has been proposed in [109] where all the received packets (intended and unintended) are exploited for iterative decoding.

The benefits of the MU MIMO-ARQ protocol can be recognized in:

- No feedback at the BS for antenna allocation.
- No feedback at the BS for user selection.

- Open-loop STBC and SM used seamlessly providing higher diversity order at low SNR regions.
- Robustness in the presence of very fast-varying channels.
- Good down-scalability: as the number of users decreases the performance is lower-bounded by the SU protocol, as in [23].

An optimized SIC-MMSE receiver is also proposed. The MT selected for hybrid SM transmission stores and exploits unintended packets sent in previous time slots to other users. The MT removes the contribution of the "overheard" packets to maximize the probability of the good reception of his packet. This receiver is later reffered to as PO.

Section 5.5 presents in detail the MU MIMO-ARQ state machine and the retransmission strategies proposed. Section 5.5.6 describes the user selection methods: the random user selection and the best user selection used to upper bound the performance achievable with MU diversity exploitation. The optimized PO receiver is explained in Section 5.6. Then, Section 5.7 reports the results before final conclusions in Section 5.8.

5.5 System Model

The DL transmission from a BS to a set of MTs – in number of N_{users} – is studied. The users are served with a RR scheduler with TDMA and fairness among users is guaranteed, as in [116]. The transmission is in broadcasting to all the users. Without loss of generality, the MIMO system is assumed with $N_T = N_R = 2$, i.e. two transmitting antenna at the BS and two receiving antenna for each MT. If more antennas are available, more than two users can be served balancing the trade-off between SM and STBC, e.g. using a layered STBC [4, 41, 49]. Open-loop MIMO schemes are used: STBC, SM and hybrid SM. In hybrid SM, two data streams are sent to different users simultaneously. An ARQ protocol is used with same symbol content and with independently coded packets. Such ARQ configuration allows to retransmit packets independently to different users in hybrid SM mode.

The packets are composed by N_{data} bits plus an appended CRC consisting of N_{crc} bits. The notation used in the design of MU MIMO-ARQ protocol is reported here:

- $X_K^{(j)}$ is the Kth packet for user j, the upper case letter X indicates the whole packet.
- $x_K^{(j)}$ is the mth symbol for user j, the lower case indicates one modulation symbol inside Kth packet.

- $\mathbf{H}^{(j)}(t)$ is the 2×2 MIMO channel matrix from the transmitter to user j at time t.
- $\mathbf{H}_i^{(j)}(t)$ is the $N_R \times 1$ channel column vector from transmit antenna i to user j at time t.

The MU MIMO-ARQ protocol transmission is started in SM mode to the first user (1) in the RR scheduler queue. If the two packets are received correctly, the BS serves the second user in the scheduler queue again in SM mode. If the two packets are received correctly, the BS transmits to the third user until the maximum number of users is reached and then it starts again from the first user. The BS schedules the packets in RR fashion.

If both packets are decoded with errors, they are retransmitted in Alamouti mode: the two transmissions (first and second time slots) constitute one Alamouti block. The Alamouti transmission is carried out until both packets are received or until the maximum number of retransmissions K_{max} is reached. In this status, the two packets are directed to the same user and the system operates in SU STBC transmission.

When one only packet is in error, the erroneous one is retransmitted coupled with a new packet. In this case, three retransmission strategies are proposed. The first strategy (S1), retransmits the packet from the best antenna with double power. This strategy needs a single bit feedback from the MT to the BS (for choosing one antenna out of two). Then, a second strategy is proposed (S2) where the packet is retransmitted from both antennas. In these first two strategies, hybrid SM is never used.

A third retransmission strategy (MU) uses hybrid SM to exploit multiuser diversity gain. It operates as follows. Assume that only one out of two packets for user (1) is received correctly. The packet in error is assumed to have been transmitted from the weakest antenna. On the first retransmission, the antenna allocation is switched. The erroneous packet is transmitted from what it is supposed to be the best antenna for communicating with user (1) and a new packet is transmitted on the other antenna for a new user (u). In RAND user selection, the new user is selected choosing randomly among all the users in the system. The performance is upper-bounded by the BEST user selection, in this case the new user is selected as the best user for throughput maximization, the user selection is detailed in Section 5.5.6. The BEST user selection needs the feedback from all the MTs to the BS of the post-processing SNRs at the two outputs of the receiver. The feedback link is assumed error-free and delay-less.

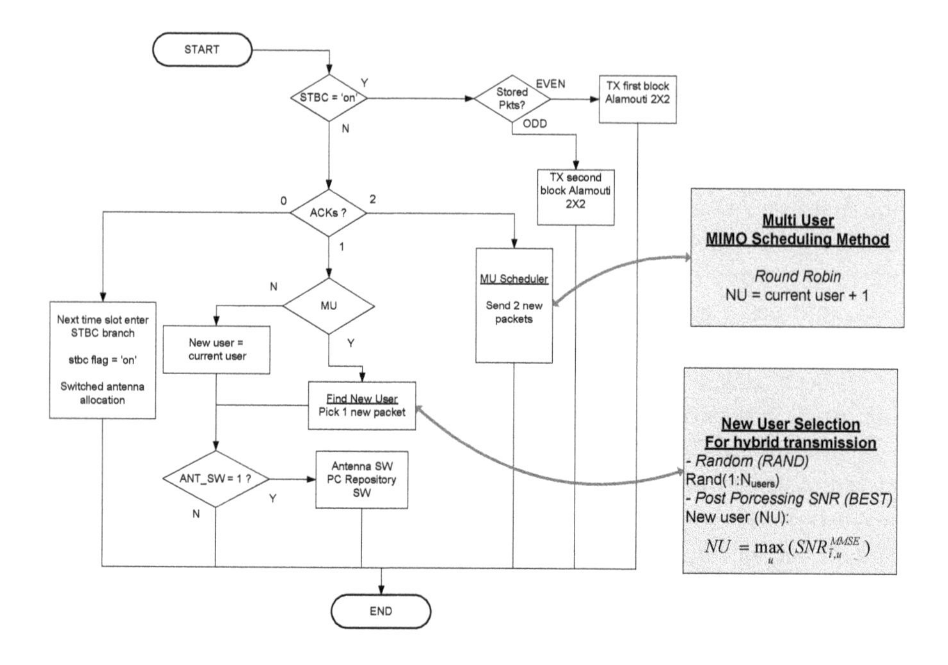

Figure 5.1 MU MIMO-ARQ Tx state machine diagram

The MIMO-ARQ protocol can be described by a state machine located at the BS which is reported in Figure 5.1. The state machine includes SU and MU systems. The main block of the state machine is the central ACKs check block. Three output conditions are possible: no packets correctly decoded (0), only one (1) or (2) correctly decoded. If both packets are not decoded $(\text{Acks} = 0)$, the antenna are switched and the STBC flag is set to "on"; in the next time slot, the transmission state machine will enter the STBC branch. In the STBC branch, if the number of Alamouti coded blocks transmitted is odd, the packets are transmitted with sign and conjugate operation as in [5]. If an even number of packets have been transmitted – i.e. an integer number of Alamouti transmissions have been completed – the packets are retransmitted unaltered to form the basis for the next Alamouti block.

If only one packet has been decoded $(\text{Acks} = 1)$, in the case of an SU system, a new packet is sent for the same user. Antenna switching criteria is evaluated at the MT and the allocation is sent back to the BS, as in [23]. In MU systems, a new user is selected (RAND/BEST). For the first retransmission, the antenna allocation is switched, while it remains fixed in the following time slots if hybrid SM transmission is still used. Hybrid SM transmission continues to be used if these two conditions are both met: the retransmitted packet has not been received and both packets are not in error. If both packets are received in error during hybrid SM mode, both packets are retransmitted in STBC. Differently from the previous case, the STBC packets are not directed to the same user and we have an MU STBC. MU STBC is terminated if the packet retransmitted for the user (1) – the scheduled user in the first time slot – is received or $K_{\max}$ retransmissions have been tried. In MU STBC, if the packet for the new user (u) is received, the system continues to retransmit both packets in STBC. This is inefficient from the user (u) prospective, but permits to preserve the information stored in the previous time slots via PC for user (1) and to preserve the diversity order of the already sent Alamouti blocks. Finally, if both packets are decoded correctly (Acks $= 2$), the scheduler transmits two new packets for the next user.

Figure 5.1 highlights the place where other schedulers and user selection strategies can be inserted in the proposed MIMO-ARQ protocol architecture. In particular, the RR scheduler can be enhanced considering QoS parameters at higher layers for the determination of the users to be scheduled. Further, the user selection process is also highlighted and other selection strategies can be inserted in the system design. In particular, a joint design of scheduling and user selection methods seems very promising for exploiting at its best the MU network resources. The methods can be designed when variable degrees of information about the MTs CSI are available at the BS. User selection method can be dynamically adapted based on the queue status – e.g. when an user has a delay approaching its maximum latency time, this user can be selected for the next hybrid SM transmission, thus giving the chance to reset his Head Of Line delay (HOL) delay.

The system seamlessly switches between STBC and SM furnishing a variable diversity order depending on the channel status. At low SNR regions, a diversity order of two is available. In the high SNR region, SM is usually used, while with moderate SNRs (from 5 to 20 dB) both schemes are used.

5.5.1 MT Receiver

The MT receiver applies Packet Combining (PC) which combines the signals received during all the retransmissions in order to have the highest probability of correct decoding the packets. The PC exploits only the packets directed to the intended user. In Section 5.6, on optimized receiver will be described

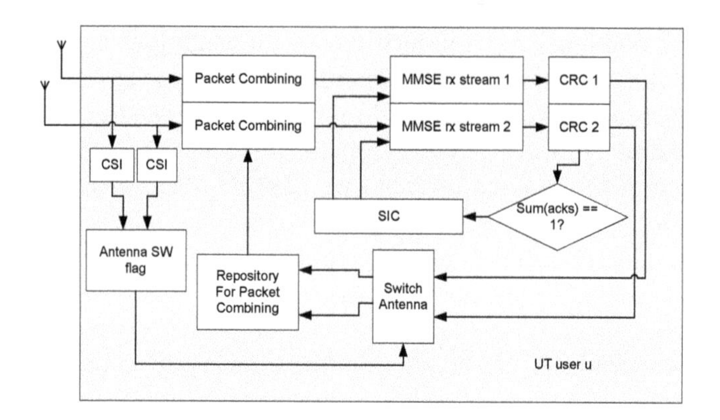

Figure 5.2 MIMO-ARQ receiver diagram

where the user not only exploits the packets directed to itself, but stores also the packets sent previously to other users.

In Figure 5.2, a receiver diagram is reported. The MT has a SIC-MMSE receiver which inverts the equivalent channel matrix $\mathbf{H}_{\text{eq}}^{(u)}$, where (u) is the user index. The data symbols are estimated via a pseudo-inversion as

$$\begin{aligned}\hat{\mathcal{X}} &= \mathcal{F}_{\text{MMSE}}^{(u)} \cdot \mathcal{Y}_u(t+k)\\ &= \left(\mathbf{H}_{\text{eq}}^{\dagger(u)}(t+k)\mathbf{H}_{\text{eq}}^{(u)}(t+k) + \frac{\sigma_n^2}{P}\right)^{-1} \mathbf{H}_{\text{eq}}^{\dagger(u)}(t) \cdot \mathcal{Y}_u(t+k)\end{aligned}$$

where $\dagger$ is the conjugate transpose operator, $\mathcal{Y}_u(t+k)$ is the vector of all the received signals until the current time slot and has dimension of $2 \times (k+1)$ and k is the number of retransmissions elapsed until the current time slot. The construction of $\mathbf{H}_{\text{eq}}^{(u)}$ depends on the protocol state evolution and it is explained in the following for each retransmission strategy.

After the MMSE channel inversion, the CRC of both streams is checked. If only one packet is decoded, the SIC part is activated: the correct packet is subtracted from the received signals and decoding is performed again. After this cancellation, the final ACKs/NACKs are communicated to the BS and the MIMO-ARQ protocol proceeds to the next transmission slot.

5.5.2 Retransmission Strategy S1

Assume that user (1) is the actual user scheduled for transmission and assume that in the first time slot only packet $X_B^{(1)}$ was correctly received and $X_A^{(1)}$ has to be retransmitted. In S1 strategy, the packet in error is retransmitted from the antenna which gives the best post-SNR at the receiver, while the other antenna is kept silent. The transmission power over the best antenna is doubled. This strategy needs one-bit feedback at the BS in order to select one antenna out of two available at the BS. It is evident that if more than two antenna are used at the BS, the feedback would need a number of bits equal to $N_b = \log_2(N_T)$.

The correctly received packet at the first time slot is cancelled and the received signals at MT_1 at time $t+1$ for the mth symbol of the packet can be written as

$$\begin{aligned}\mathcal{Y}_1 &= \begin{bmatrix} \tilde{y}_1(t) \\ y_1(t+1) \end{bmatrix} \\ &= \begin{bmatrix} \mathbf{H}_1^{(1)}(t) \\ \mathbf{H}_B^{(1)}(t+1) \end{bmatrix} \cdot x_A^{(1)} + \mathcal{N}\end{aligned}$$

where $\mathbf{H}_B^{(1)}(t+1)$ is the channel column vector for the antenna with the best post-processing SNR for user (1), $\mathcal{N} = (n_A(t) \quad n_B(t))^T$ is the Gaussian noise term, $n_i(t)$ is $\mathcal{CN}(0, 1)$ and the cancelled signals is

$$\hat{y}_1(t) = y_1 - \mathbf{H}_1^{(1)}(t) \cdot x_B^{(1)} \tag{5.5}$$

If the packet is still not received, at the second time slot, the received signal is expressed as

$$\begin{aligned}\mathcal{Y}_1 &= \begin{bmatrix} \tilde{y}_1(t) \\ y_1(t+1) \\ y_1(t+2) \end{bmatrix} \\ &= \begin{bmatrix} \mathbf{H}_1^{(1)}(t) \\ \mathbf{H}_B^{(1)}(t+1) \\ \mathbf{H}_B^{(1)}(t+2) \end{bmatrix} \cdot x_A^{(1)} + \mathcal{N}\end{aligned}$$

This process holds until the packet is received or the maximum number of retransmission $K_{\max}$ has not been reached.

5.5.3 Retransmission Strategy S2

Assume that user (1) is the actual user scheduled for transmission and assume that in the first time slot only packet $X_B^{(1)}$ was correctly received and $X_A^{(1)}$ has to be retransmitted. In S2 strategy, the packet in error is retransmitted from both antenna in spatial repetition coding. The MT does not need to send the feedback of the preferred antenna to the BS, since both antenna are used.

In strategy S2, after packet $X_B^{(1)}$ has been cancelled in the first time slot, the received signal at the second time slot for the mth symbol of the packet is expressed as

$$\begin{aligned}\mathcal{Y}_1 &= \begin{bmatrix} \tilde{y}_1(t) \\ y_1(t+1) \end{bmatrix} \\ &= \begin{bmatrix} \mathbf{H}_1^{(1)}(t) \\ \mathbf{H}_1^{(1)}(t+1) + \mathbf{H}_2^{(1)}(t+1) \end{bmatrix} \cdot x_A^{(1)} + \mathcal{N}\end{aligned}$$

The retransmission strategy S2 transmits the same signal from the two antenna and do not doubles the power at the antenna port. The power doubling required by strategy S1 is correct from an information theoretical perspective but practical problems would be encountered if S1 is applied in real system. The operating point of the RF transmitter chain will not tolerate a power totally unbalanced on a single antenna element. For this reason, S2 strategy seems more appropriate for a practical system design. The act of transmitting from both antennas can be regarded as a form of repetition coding: *spatial repetition coding*.

5.5.4 Retransmission Strategy MU

The third proposed strategy MU uses hybrid SM transmission in order to exploit multi-user diversity. This strategy leads to a more complex MIMO-ARQ state machine and several possible states are possible.

Assume that user (1) is scheduled in the first time slot, strategy MU is illustrated for the following time evolution which summarizes all the configurations possible:

- First time slot t: transmission of packets $X_A^{(1)}$ and $X_B^{(1)}$ to user (1), only packet $X_A^{(1)}$ is decoded.
- $t+1$: retransmission of $X_B^{(1)}$ from switched antenna; user $(u+1)$ is selected for hybrid SM transmission of packet $X_A^{(u+1)}$. Packet $X_B^{(1)}$ still not decoded, while packet $X_A^{(u+1)}$ is correctly decoded at MT_{u+1}.
- $t+2$: retransmission of $X_B^{(1)}$ (antenna is not anymore switched); user $(u+2)$ is selected for hybrid SM transmission of packet $X_A^{(u+2)}$. Packet X_B^1 is not decoded at MT_1 and $X_A^{(u+2)}$ is not decoded at MT_{u+2}.

Table 5.1 MU MIMO-ARQ: case study

Time Slot	Tx Ant 1	Dec	Tx Ant2	Dec
t	$X_A^{(1)}$	$\surd$	$X_B^{(1)}$	$\times$
$t+1$	$X_B^{(1)}$	$\times$	$X_A^{(u+1)}$	$\surd$
$t+2$	$X_B^{(1)}$	$\times$	$X_A^{(u+2)}$	$\times$
$t+3$	$-X_A^{\dagger(u+2)}$	$\times$	$X_B^{\dagger(1)}$	$\times$
$t+4$	$X_B^{(1)}$	$\surd$	$X_A^{(u+2)}$	$\surd$

- $t+3$: retransmission in MU STBC with Alamouti 2×2; both packets are not decoded.
- $t+4$: retransmission of the first half of a new Alamouti block; both packets are decoded correctly; the scheduler is reset and packets $X_A^{(2)}$ and $X_B^{(2)}$ are transmitted for the second user in the scheduler queue.

Table 5.1 reports the packets sent on each antenna at the BS with the relative CRC checking outcome at the intended MTs for each time slot.

In the following, the received signals at the MTs are expressed along with the construction of the equivalent channel matrix: $\mathbf{H}_{\text{eq}}^{(u)}$. In the first time slot, user (1) receives packets $X_A^{(1)}$ and $X_B^{(1)}$ in SM at time t. The corresponding received signal for the mth symbol of the packet is

$$y_1(t) = \mathbf{H}^{(1)}(t) \cdot \begin{pmatrix} x_A^{(1)} \\ x_B^{(1)} \end{pmatrix} + \mathcal{N}(t)$$

where $\mathbf{H}^{(1)}(t) = [\mathbf{H}_1^{(1)}(t)|\mathbf{H}_2^{(1)}(t)]$.

Assume that packet X_A^1 is decoded. The SIC is performed and the contribution from packet $X_A^{(1)}$ is removed: the interference-cancelled signal is

$$\tilde{y}_1(t) \;=\; y_1 - \mathbf{H}_1^{(1)}(t) \cdot x_A^{(1)} \tag{5.6}$$

After the cancellation, packet $X_B^{(1)}$ is decoded again, if not successful in the next time slot the packet X_B^1 is transmitted from switched antenna (i.e. one) and the packet X_A^{u+1} for a new user is sent over antenna two. The received signal at the MT_1 can be written including the previous cancelled received signal $\hat{y}_1$ and the new time slot transmission:

$$\begin{aligned}\mathcal{Y}_1 &= \begin{bmatrix} \tilde{y}_1(t) \\ y_1(t+1) \end{bmatrix} \\ &= \begin{bmatrix} \mathbf{H}_2^{(1)}(t) & \mathrm{O}_{2\times 1} \\ \mathbf{H}_1^{(1)}(t+1) & \mathbf{H}_2^{(1)}(t+1) \end{bmatrix} \begin{bmatrix} x_B^{(1)} \\ x_A^{(u+1)} \end{bmatrix} + \mathcal{N}(t)\end{aligned}$$

where $\mathcal{N}(t)$ is a four-term additive Gaussian noise for $\mathcal{Y}_1$.

The received signal at the MT_{u+1} is composed by the two terms, one for each antenna for the current $t+1$ time slot:

$$y_{u+1}(t+1) = \mathbf{H}^{(u+1)}(t+1) \cdot \begin{pmatrix} x_B^{(1)} \\ x_A^{(u+1)} \end{pmatrix} + \mathcal{N}(t) \tag{5.7}$$

where $\mathbf{H}^{(u+1)}(t+1) = [\mathbf{H}_1^{(u+1)}(t+1) \mid \mathbf{H}_2^{(u+1)}(t+1)]$.

Now suppose that packet $X_B^{(1)}$ is not correctly decoded and the new packet in hybrid transmission $X_A^{(u+1)}$ is decoded at MT_{u+1}. In the third time slot, packet $X_B^{(1)}$ is retransmitted without antenna switching and a new packet $X_A^{(u+2)}$ is sent on the second antenna for a new user $(u+2)$. At MT_1 the signal is thus composed as

$$\begin{aligned}\mathcal{Y}_1 &= \begin{bmatrix} \tilde{y}_1(t) \\ y_1(t+1) \\ y_1(t+2) \end{bmatrix} \\ &= \begin{bmatrix} \mathbf{H}_2^{(1)}(t) & \mathrm{O}_{2\times 1} & \mathrm{O}_{2\times 1} \\ \mathbf{H}_1^{(1)}(t+1) & \mathbf{H}_2^{(1)}(t+1) & \mathrm{O}_{2\times 1} \\ \mathbf{H}_1^{(1)}(t+2) & \mathrm{O}_{2\times 1} & \mathbf{H}_2^{(1)}(t+2) \end{bmatrix} \cdot \begin{bmatrix} x_B^{(1)} \\ x_A^{(u+1)} \\ x_A^{(u+2)} \end{bmatrix} + \mathcal{N}\end{aligned}$$

The received signal at MT_{u+2} is

$$y_{u+2}(t+2) = \mathbf{H}^{(u+2)}(t+2) \cdot \begin{pmatrix} x_B^{(1)} \\ x_A^{(u+2)} \end{pmatrix} + \mathcal{N}(t) \tag{5.8}$$

where $\mathbf{H}^{(u+2)}(t+2) = [\mathbf{H}_1^{(u+2)}(t+2) \mid \mathbf{H}_2^{(u+2)}(t+2)]$. Now assume that both packets at MT_1 and MT_{u+2} are decoded with errors. An MU-STBC transmission is started with a 2×2 Alamouti scheme. The STBC transmission ends only when both packets are decoded correctly at the MT_1 or $K_{\max}$ retransmission are achieved. In the next time slot, at MT_1, the received signal is expressed as

$$\mathcal{Y}_1 = \begin{bmatrix} \tilde{y}_1(t) \\ y_1(t+1) \\ y_1(t+2) \\ y_1^\dagger(t+3) \end{bmatrix} = \begin{bmatrix} \mathbf{H}_2^{(1)}(t) & O_{2\times 1} & O_{2\times 1} \\ \mathbf{H}_1^{(1)}(t+1) & \mathbf{H}_2^{(1)}(t+1) & O_{2\times 1} \\ \mathbf{H}_1^{(1)}(t+2) & O_{2\times 1} & \mathbf{H}_2^{(1)}(t+2) \\ \mathbf{H}_2^{\dagger(1)}(t+3) & O_{2\times 1} & -\mathbf{H}_1^{\dagger(1)}(t+3) \end{bmatrix} \cdot \begin{bmatrix} x_B^{(1)} \\ x_A^{(u+1)} \\ x_A^{(u+2)} \end{bmatrix} + \mathcal{N} \tag{5.9}$$

At MT_{u+2} the signal is composed by a full STBC block as

$$\mathcal{Y}_{u+2} = \begin{bmatrix} y_{u+2}(t+2) \\ y_{u+2}^\dagger(t+3) \end{bmatrix} = \begin{bmatrix} \mathbf{H}_1^{(u+2)}(t+2) & \mathbf{H}_2^{(u+2)}(t+2) \\ \mathbf{H}_2^{\dagger(u+2)}(t+3) & -\mathbf{H}_1^{\dagger(u+2)}(t+3) \end{bmatrix} \cdot \begin{pmatrix} x_B^{(1)} \\ x_A^{(u+2)} \end{pmatrix} \tag{5.10}$$

The expression of the equivalent channel matrix to be inverted by the MMSE requires that the signal received for the second half of the STBC block is conjugated, as expressed with $y_1^\dagger(t+3)$ and $y_{u+2}^\dagger(t+3)$ in (5.9) and (5.10) respectively. If the packets are not correctly decoded, a new first half of STBC block is transmitted. For brevity, the received signal is reported only for MT_1 (note that the received signal for the first half of the new STBC block is not conjugated):

$$\mathcal{Y}_1 = \begin{bmatrix} \tilde{y}_1(t) \\ y_1(t+1) \\ y_1(t+2) \\ y_1^\dagger(t+3) \\ y_1(t+4) \end{bmatrix} = \begin{bmatrix} \mathbf{H}_2^{(1)}(t) & O_{2\times 1} & O_{2\times 1} \\ \mathbf{H}_1^{(1)}(t+1) & \mathbf{H}_2^{(1)}(t+1) & O_{2\times 1} \\ \mathbf{H}_1^{(1)}(t+2) & O_{2\times 1} & \mathbf{H}_2^{(1)}(t+2) \\ \mathbf{H}_2^{\dagger(1)}(t+3) & O_{2\times 1} & -\mathbf{H}_1^{\dagger(1)}(t+3) \\ \mathbf{H}_1^{(1)}(t+4) & O_{2\times 1} & \mathbf{H}_2^{(1)}(t+4) \end{bmatrix} \cdot \begin{bmatrix} x_B^{(1)} \\ x_A^{(u+1)} \\ x_A^{(u+2)} \end{bmatrix} + \mathcal{N} \tag{5.11}$$

Once the STBC transmission is started, the scheduler can proceed to the next user in the queue only when:

- the packet $X_B^{(1)}$ at MT_1 is decoded,
- maximum number of retransmission has been reached (K_{max}).

5.5.5 SIC Principle

To illustrate the SIC principle and how it is applied throughout the protocol operation, let us focus on MT_1. Beside the cancellation of packet X_A^1 (or X_B^1) in the first time slot, SIC can be applied at each step of the protocol evolution. The vector $\hat{\mathcal{X}}$ contains the estimates of the packet(s) to be received at the current MT and also other packets directed to other users, if hybrid SM transmission mode has been used. Suppose that the intended packet is not received. If one (or more) of the other packets is/are correctly decoded, the MT cancels their contributions and then tries to recover again the remaining packets. This can be done several times if a new packet is decoded successful at each SIC iteration.

To illustrate this mechanism, let us refer to (5.11). At the output of the receiver, three estimates are available: $[\hat{x}_B^{(1)} \quad \hat{x}_A^{(u+1)} \quad \hat{x}_A^{(u+2)}]^T$. If $X_B^{(1)}$ packet is decoded correctly the scheduler can serve the next user in the queue. Otherwise, $X_A^{(u+1)}$ and/or $X_A^{(u+2)}$ can be decoded correctly. The correctly received packets are subtracted from the received signal vector and MMSE is performed again. If, say, $X_A^{(u+1)}$ was decoded correctly, a cancelled signal is constructed as

$$\tilde{\mathcal{Y}}_1 = \mathcal{Y}_1 - \mathbf{H}_{eq,2} \cdot X_A^{(u+1)} \tag{5.12}$$

where $\mathbf{H}_{eq,2} = [\mathrm{O}_{2\times 1} \quad \mathbf{H}_2^{(1)}(t+1) \quad \mathrm{O}_{6\times 1}]^T$ is the second column of the channel equivalent matrix. The estimates are then obtained via the MMSE inversion:

$$\hat{\mathcal{X}}_1 = \mathcal{F}_{\text{MMSE}}, \quad \tilde{\mathcal{Y}}_1 = \left(\mathbf{H}_{eq}^{\dagger}\mathbf{H}_{eq} + \frac{\sigma_n^2}{P}\right)^{-1} \mathbf{H}_{eq}^{\dagger}, \quad \tilde{\mathcal{Y}}_1 \tag{5.13}$$

If, at this step, $X_B^{(u+2)}$ was decoded without errors, a new SIC step could have been enforced for the last chance of detecting $X_B^{(1)}$:

$$\tilde{\tilde{\mathcal{Y}}}_1 = \tilde{\mathcal{Y}}_1 - \mathbf{H}_{eq,3} \cdot X_A^{(u+2)} \tag{5.14}$$

where the third column of the equivalent channel matrix is

$$\mathbf{H}_{eq,3} = \left[\mathrm{O}_{4\times 1} \quad \mathbf{H}_2^{(1)}(t+2) \quad -\mathbf{H}_1^{(1)}(t+3) \quad \mathbf{H}_2^{(1)}(t+4)\right]^T$$

5.5.6 New User Selection

With the MU strategy, hybrid SM mode is used and a new user is coupled with the previous scheduled user which needs a single packet retransmission. On one hand, the new user selection can be performed without any knowledge at the transmitter when the BS chooses randomly the new user to be scheduled. On the other hand, if every MT reports to the BS the post-processing SNRs at the outputs of the MMSE receiver for each stream, the BS can chose the best user in order to maximize the system throughput. Since the BEST user selection needs feedback of the post-processing SNR from each MT to the BS, this strategy is here proposed as a comparison.

The BEST user selection is reliable when the channel between successive time slots are highly time-correlated. The issue of the feedback rate given the variability of the MT CSI is not in the scope of this work and the best user selection is performed assuming that $\mathbf{H}^P(t+1) = \mathbf{H}(t)$, i.e. the channels in the future time slot are assumed equal to the channels in the current time slot and the time dependency is dropped in the following.

5.5.6.1 MMSE Receiver and Post-Processing SNR

In this section, the post-processing for the MMSE is expressed: the estimation error for the mth symbol for the Kth packet directed to the user u can be expressed as

$$
\begin{aligned}
e^{\text{MMSE}} &= \min[E(x_K^{(u)} - \hat{x}_K^{(u)})(x_K^{(u)} - \hat{x}_K^{(u)})^\dagger] \\
&= \sigma_n^2 \left(\mathbf{H}^{(u)^\dagger} \mathbf{H}^{(u)} + \frac{\sigma_n^2}{P} \right)^{-1}
\end{aligned} \tag{5.15}
$$

where $\hat{x}_K^{(u)}$ is the MMSE estimation of x_K directed to the user (u), $K \in \{A, B\}$, σ_n^2 is the noise variance, P is the power transmitted for each stream and $\mathbf{H}^{(u)}$ is the equivalent channel matrix for user (u). The post-processing SNR for each stream i thus can be expressed as

$$
\text{SNR}_i^{\text{MMSE}} = \frac{P}{e^{\text{MMSE}}} = \frac{P}{\left[\sigma_n^2 \left(\mathbf{H}^{(u)^\dagger} \mathbf{H}^{(u)} + \frac{\sigma_n^2}{P} \right)^{-1} \right]} \tag{5.16}
$$

Expanding the matrix notation in (5.16) for the 2×2 MIMO case, the post-processing can be expressed as

$$
\text{SNR}_i^{\text{MMSE}} = \frac{P}{\sigma_n^2} \left[\left(\|\mathbf{H}_i^{(u)}\|^2 + \frac{\sigma_n^2}{P} \right) - \frac{|\mathbf{H}_i^{(u)^\dagger} \mathbf{H}_{\underline{i}}^{(u)}|^2}{\|\mathbf{H}_{\underline{i}}\|^2 + \frac{\sigma_n^2}{P}} \right] \tag{5.17}
$$

where the alternative antenna to the ith is indicated with

$$\underline{i} = 3 - i \quad \text{with} \quad i = \{1, 2\}$$

The post-processing SNR for the ith stream depends on $\mathbf{H}_i^{(u)}$, which is the channel from the two transmitting antenna to the ith receiving one, but it is also degraded from the non-perfect orthogonality between the $\mathbf{H}_i^{(u)}$ and $\mathbf{H}_{\underline{i}}^{(u)}$ as represented in the second subtractive term.

When the same modulation is used for both streams [23], the best user selection for throughput maximization is to select the user with the best post-processing SNR. Each user calculates the two post-processing SNR at the output of the receiver ($\text{SNR}_1^{\text{MMSE}}$ and $\text{SNR}_2^{\text{MMSE}}$) and sends this information back to the BS. The BS has all the information needed to optimize the multi-user scheduling, the New User (NU) is decided searching the N_{users} space:

$$\text{NU} = \max_u \left(\text{SNR}_{\underline{i},u}^{\text{MMSE}}\right) \tag{5.18}$$

where $\text{SNR}_{\underline{i},u}^{\text{MMSE}}$ is the post-processing SNR for user u for the antenna $\underline{i}$ which will transmit the new packet. The best user selection is performed after the antenna switching criterion has been evaluated.

5.6 PO – Packet Overhearing

The performance of the system can be further improved with an optimized design of the SIC-MMSE receiver. When hybrid SM transmission is used, one antenna retransmits a previous packet which was decoded incorrectly jointly with a packet directed to a new user. The probability that the retransmitted packet will be decoded (due to the packet combining at the receiver) can be reasonably high. The probability of decoding the new packet, instead, will be lower since it is based on a single received signal collected at the new MT.

An optimized MT receiver with the ability of recording previous packets transmitted is described here. The packet storing starts when the MIMO-ARQ protocol is reset and a new user is scheduled in the system. This condition is notificated by the BS. The MIMO-ARQ protocol can be reset when the two packets directed to the scheduled user have been decoded or when the maximum number of retransmission has been reached.

The PO receiver has an impact on the system performance due to a two-fold improvement. Firstly, the joint detection at the MMSE receiver of two (or more) received signals improves the probability of decoding the new packet sent in the hybrid SM transmission. Secondly, the SIC procedure of the PO receiver has more possibilities due to the larger set of estimates.

5.6.1 Received Signals at the Second Time Slot

Let us consider a transmission to MT_1 with packets $X_A^{(u)}$ and $X_B^{(u)}$. If MT_1 has decoded correctly both packets, the BS skips to the next user in the scheduling queue and user $(u+1)$ deletes the overheard signals from its memory. If MT_1 decodes only one packet, say $X_A^{(1)}$, $X_B^{(1)}$ is retransmitted with switched antenna allocation (as described in Section 5.5) and a new packet $X_A^{(u+1)}$ is sent for the new user $(u+1)$.

The aim of user $(u+1)$ is to detect the packet $X_A^{(u+1)}$ exploiting all the radio signals collected. In the next time slot, two cases can occur at MT_1:

1. MT_1 has decoded $X_B^{(1)}$ after one retransmission,
2. MT_1 has decoded $X_B^{(1)}$ incorrectly.

In the first case, the hybrid SM transmission is terminated and the MU scheduler will compose two new packets for the next user in the scheduler queue. The fact that MT_{u+1} decodes or not its packet is not affecting the protocol state evolution; the protocol is reset anyway. In the second case where MT_1 has decoded the packet with errors, $X_B^{(1)}$ is retransmitted from the same antenna. If MT_{u+1} has decoded its packet, a new user will be scheduled in hybrid SM, otherwise a Multi-User STBC is started for packets $X_A^{(1)}$ and $X_B^{(u+1)}$.

In either cases, at the end of the second time slot transmission, user $(u+1)$ has, so far, collected four radio signals. In the first time slot packets $X_A^{(1)}$ and $X_B^{(1)}$; in the second time slot the packet $X_B^{(1)}$ (retransmitted to user 1) and $X_A^{(u+1)}$ transmitted for the first time. At MT_{u+1}, each signal is transmitted on channel $\mathbf{H}^{(u+1)}$ as follows:

- $X_A^{(u)}$: for user 1 from antenna 1. Corresponding (noise free) received signal for the mth symbol: $\mathbf{H}_1^{(u+1)}(t)x_A^{(1)}$.
- $X_B^{(u)}$: for user 1 from antenna 2. Corresponding (noise free) received signal for the mth symbol: $\mathbf{H}_2^{(u+1)}(t)x_B^{(1)}$.
- $X_B^{(u)}$: for user u retransmitted from antenna 1. Corresponding (noise free) received signal for the mth symbol: $\mathbf{H}_1^{(u+1)}(t+1)x_B^{(u+1)}$.
- $X_A^{(u+1)}$: for user $u+1$ from antenna 2. Corresponding (noise free) received signal for the mth symbol: $\mathbf{H}_2^{(u+1)}(t+1)x_A^{(u+1)}$.

The received signal at MT_{u+1} for the first time slot can be expressed as

$$y_{u+1}(t) = \mathbf{H}_1^{(u+1)}(t)x_A^{(u)} + \mathbf{H}_2^{(u+1)}(t)x_B^{(u)} + \mathcal{N}^{(u+1)}(t) \tag{5.19}$$

where $\mathbf{H}_i^{(u+1)}(t)$ is the channel from transmitting antenna i to MT_{u+1} at time t. The received signal during the retransmission $(t+1)$ is

$$y_{u+1}(t+1) = \mathbf{H}_1^{(u+1)}(t+1)x_B^{(u)} + \mathbf{H}_2^{(u+1)}(t+1)x_A^{(u+1)} + \mathcal{N}^{(u+1)}(t+1) \tag{5.20}$$

The received signals can be composed for a single MMSE channel inversion to maximize the probability of correctly decoding packet $X_A^{(u+1)}$:

$$\begin{aligned}
\mathcal{Y} &= \begin{bmatrix} \mathbf{H}_1^{(u+1)}(t) \\ \mathrm{O}_{2\times1} \end{bmatrix} x_A^{(u)} + \begin{bmatrix} \mathbf{H}_2^{(u+1)}(t) \\ \mathbf{H}_1^{(u+1)}(t+1) \end{bmatrix} x_B^{(u)} \\
&\quad + \begin{bmatrix} \mathrm{O}_{2\times1} \\ \mathbf{H}_2^{(u+1)}(t+1) \end{bmatrix} x_A^{(u+1)} + \begin{bmatrix} \mathcal{N}^{(u)}(t) \\ \mathcal{N}^{(u)}(t+1) \end{bmatrix} \\
&= \begin{bmatrix} \mathbf{H}_1^{(u+1)}(t) & \mathbf{H}_2^{(u+1)}(t) & \mathrm{O}_{2\times1} \\ \mathrm{O}_{2\times1} & \mathbf{H}_1^{(u+1)}(t+1) & \mathbf{H}_2^{(u+1)}(t+1) \end{bmatrix} \begin{bmatrix} x_A^{(u)} \\ x_B^{(u)} \\ x_A^{(u+1)} \end{bmatrix} \\
&= \mathbf{H}_{4\times3} \cdot \mathcal{X}
\end{aligned} \tag{5.21}$$

Then, the MMSE estimate is obtained via multiplication with $\mathcal{F}_{\mathrm{MMSE}}$ which is the pseudo-inverse of $\mathbf{H}_{4\times3}$:

$$\hat{\mathcal{X}} = \mathcal{F}_{\mathrm{MMSE}}, \quad \mathcal{Y} = \left(\mathbf{H}_{4\times3}^{\dagger}\mathbf{H}_{4\times3} + \frac{\sigma_n^2}{P}\right)^{-1} \mathbf{H}_{4\times3}^{\dagger}, \quad \mathcal{Y} \tag{5.22}$$

After MMSE receiver, if packet $X_A^{(u+1)}$ is not received, there is a double SIC possibility: packet X_A^1 and/or packet X_B^1 can be subtracted if correctly decoded. If PO was not used, only packet X_A^1 could have been exploited for SIC.

5.6.2 Generalization of PO for Any Time Slot

The PO is possible at any step of the MIMO-ARQ protocol operation, i.e. for any user chosen for hybrid SM transmission at any time slot. To exemplify this case, let us present the most general case where user $(u+2)$ is selected for hybrid SM transmission at time slot $t+2$.

Assume that packets $X_A^{(1)}$ and packets $X_B^{(1)}$ are transmitted in the first time slot for MT_1 and assume that packet $X_A^{(1)}$ has not been decoded. Packet $X_A^{(1)}$

is retransmitted with switched antenna. A first user $(u + 1)$ is scheduled in hybrid SM together with packet $X_A^{(1)}$. MT_{u+1} decoded correctly its packet.

At the successive time slot, user $(u+2)$ is scheduled. MT_1 and MT_{u+2} are not able to decode the packets and a multi-user STBC starts. After an Alamouti coded block, both packets are decoded and the MIMO-ARQ protocol can be finally reset without excess of retransmissions.

The received signals at MT_{u+2}, after reception of a full STBC block (fourth time slot), can be expressed as reported in (5.23):

$$
\begin{aligned}
\mathcal{Y}_{u+2}(t+3) &= \begin{bmatrix} Y_{u+2}(t) \\ Y_{u+2}(t+1) \\ Y_{u+2}(t+2) \\ Y^{\dagger}_{u+2}(t+3) \end{bmatrix} \\
&= \begin{bmatrix} \mathbf{H}_1^{(u+2)}(t) & \mathbf{H}_2^{(u+2)}(t) & O_{2\times 1} & O_{2\times 1} \\ O_{2\times 1} & \mathbf{H}_1^{(u+2)}(t+1) & \mathbf{H}_2^{(u+2)}(t+1) & O_{2\times 1} \\ O_{2\times 1} & \mathbf{H}_1^{(u+2)}(t+2) & O_{2\times 1} & \mathbf{H}_2^{(u+2)}(t+2) \\ O_{2\times 1} & -\mathbf{H}_2^{\dagger(u+2)}(t+3) & O_{2\times 1} & \mathbf{H}_1^{\dagger(u+2)}(t+3) \end{bmatrix} \\
&= \begin{bmatrix} x_A^{(1)} \\ x_B^{(1)} \\ x_A^{(u+1)} \\ x_A^{(u+2)} \end{bmatrix} + \mathcal{N}
\end{aligned}
\tag{5.23}
$$

5.7 Results

In this section, results are reported for different MIMO-ARQ protocol configurations including PO receivers. The protocol configurations are listed for clarity in Table 5.2 and include SU and MU modes. The performance of the proposed strategies are compared with a system with non-adaptive MIMO schemes and in the case when the schemes are selected based on a maximum BER, as described in Chapter 3.

The PHY layer is configured as reported in Table 5.3. The channel is simulated in block fading. The channel power is normalized to one. Where not otherwise specified, each fading block contains 250 packets, which means a channel coherence time of $\approx$ 108 ms. The total number of packets simulated is 25000.

Table 5.2 MU MIMO-ARQ: configurations

Protocol Configuration	SU/MU	User Selection
SU	SU as in [23]	–
S1	SU	–
S2	SU	–
MU RAND	Hybrid MU	Random
MU BEST	Hybrid MU	Post-proc SNR
MU RAND PO	Hybrid MU PO	Random
MU BEST PO	Hybrid MU PO	Post-proc SNR

Table 5.3 System parameters

Parameter	Value
Modulation	4 QAM
Symbol duration	1 μs
Symbols per packet	432
Packet duration	$T_p = 432\ \mu$s
Length of CRC	17 bit
Data bits per packet	847
Number of users	10
Max number of rtx ($K_{\max}$)	7
Channel	Rayleigh fading
Total packets simulated per SNR	25000
Block fading length in packets	250

Results are reported in terms of system throughput and HOL delays. The RR scheduler provides fairness among all the users and the results are presented averaged over the users.

5.7.1 Cross-Layer MIMO-ARQ versus Single-Layer Adaptivity

The average rate and delay of the MU strategy are compared with three systems: fixed STBC, fixed SM and with a system where the two former schemes are selected based on a maximum BER at the PHY layer as in Chapter 3 [74].

The fixed STBC system transmits Alamouti blocks; the CRC is checked only after an even number of transmission slots have been received (i.e. after a full Alamouti block has been received). In the fixed SM system, if two packets are in error, they are retransmitted as are. If one packet is in error the packet is retransmitted from the same antenna and a new packet is scheduled on the other antenna.

The adaptive system switches from the fixed STBC to the fixed SM transmission when the SNR is high enough to support SM transmission with a BER lower than a given threshold. From STBC to SM, the spectral efficiency is doubled. Figure 5.3 plots the BER curve for 4-QAM Alamouti and 4-QAM VBLAST schemes which is used for the determination of the cross-over point between SM and STBC. Two BER thresholds are chosen in the following: BER ≤ 0.01 and BER ≤ 0.03. In all the three systems (adaptive, fixed STBC and fixed SM), the receiver is a SIC-MMSE with Packet Combining.

Figure 5.4 reports the rate averaged over all the users for a fixed STBC and a fixed SM system, the rate for an adaptive scheme (with two BER thresholds) and the rate achieved by the MU strategy with PO in the case of BEST and RAND user selection. The adaptive system chooses the best scheme depending on the channel status and clearly outperforms the fixed MIMO systems.

The adpative system is outperformed by the proposed MU strategy: the faster adaptivity of the MIMO-ARQ state machine offers higher data rates and lower average delays due to added flexibility in the receiver and exploitation of the multi-user diversity.

As evident in Figure 5.4, the low SNR region is dominated by the diversity gain of the STBC. The fixed STBC system data rate is converging to almost 200 Kbps when SNR $\rightarrow \infty$, while fixed SM reaches 400 Kbps: as expected due to the double SE. The different slopes of STBC and SM in the range of [0, 5] dB represent the double and single diversity of the SM and STBC schemes respectively. When the BER is constrained to be lower than 0.01, the system switches to SM after 12.5 dB. In the case of a 0.03 BER threshold, the system switches after 10 dB.

The MU strategy offers a better performance over all the SNR range. In the low SNR region, the gain of the MU in respect to the fixed STBC system comes from the possibility of receiving also halves of the STBC transmission blocks (as explained in Section 5.5.4) and from the fact that sometimes, in the case of particularly good CSI, SM can be used with an overall capacity gain. Around a 1 dB gain can be observed when SNR $= 0$ dB. MU strategy and fixed STBC offer same data rate when $5 \leq \text{SNR} \leq 7.5$ dB: in this SNR region the higher probability of having a single packet in error during the first time slot pushes the MU strategy in longer retransmission states, compared to the fixed STBC system. After 7.5 dB, the MU strategy offers almost 5 dB increase over the fixed SM scheme. The gain comes from exploitation of multi-user diversity and from the seamlessly usage of STBC and SM: in the high SNR region, still STBC can be seldom used when the users are occasionally in bad fading.

In Figures 5.5 and 5.6, the average delays and the average delay jitters are plotted. In fixed STBC, the delay is converging to around 8 ms against the

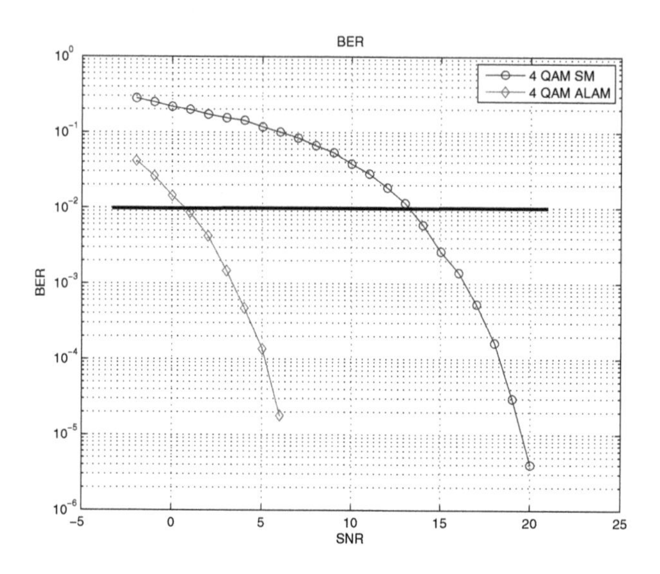

Figure 5.3 4-QAM Alamouti and VBLAST BERs over SNR

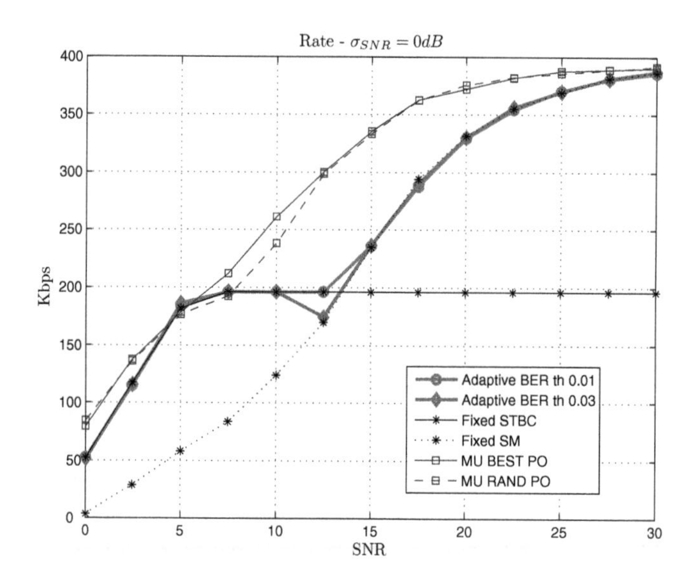

Figure 5.4 Average rate for MU MIMO-ARQ compared with non-adaptive and BER-adaptive systems

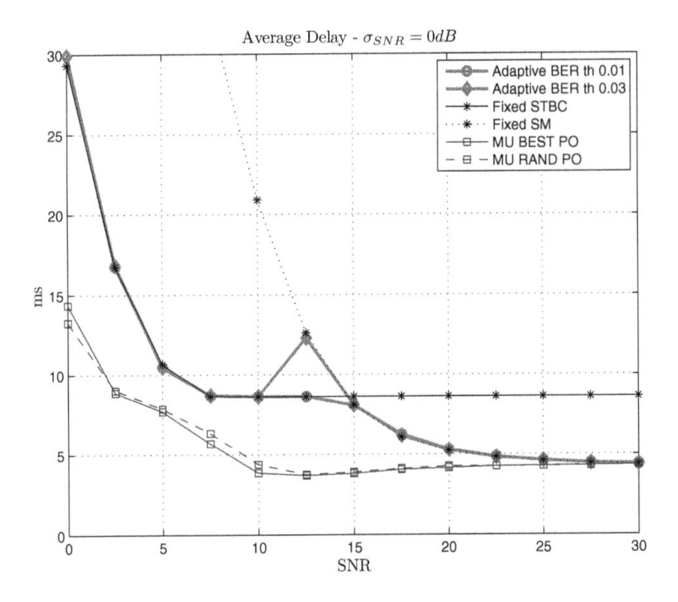

Figure 5.5 Average delay for MU MIMO-ARQ compared with non-adaptive and BER-adaptive systems

4 ms of the fixed SM scheme because a full Alamouti block has to be received before decoding. The adpative schemes present a delay which stabilizes at 8 ms for SNR = 10 dB. Thereafter, the delay lowers to the level of the fixed SM system.[1]

The MU strategy sensibly improves the delay performance due to the hybrid SM transmissions: when one packet is in error, it is still possible to chose another user to send a packet to. In the case of correct reception the new user chosen for hybrid SM has a reduction of his average delay. This results in the fact that MU strategy approaches the 4 ms average delay already at around 10 dB with a considerably lower delay in the low SNR region. At 5 dB the average delay is 2 ms lower than in the STBC system, i.e. the users are served in average four time slots earlier than in the other three systems.

[1] The average delay is converging to 4.32 ms when SNR $\rightarrow \infty$: indeed all the packets are transmitted in SM and successfully decoded. With RR scheduler, each MT receives two packets every $1/N_{\text{users}} \cdot T_p$ and the asymptotic average delay depends on the number of users in the scheduler queue. If other schedulers are integrated in the system, the asymptotic delay for each user may be different.

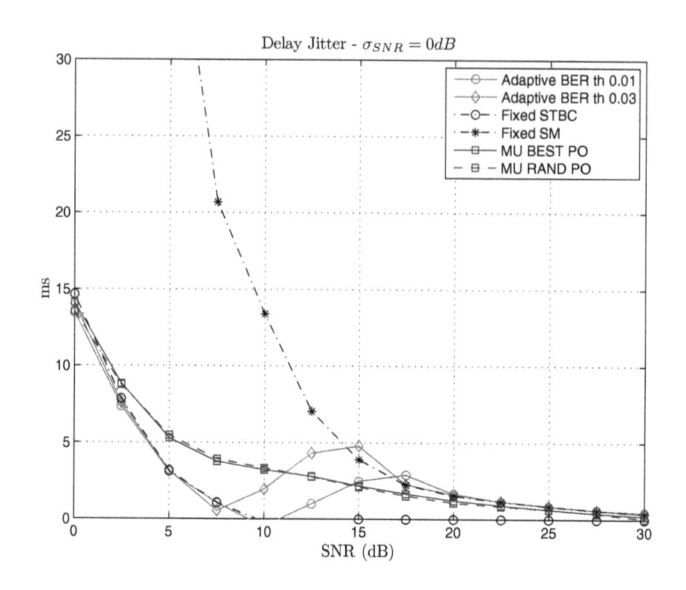

Figure 5.6 Delay jitter for MU MIMO-ARQ compared with non-adaptive and BER-adaptive systems

The delay jitter of the MU strategy is higher than in the fixed STBC strategy but it is lower-bounding the one of the fixed SM scheme. The higher delay jitter is due to the larger set of states available in the MIMO-ARQ protocol compared to the unique retransmission possibility for the fixed STBC.

5.7.2 S1 – S2 – MU – MU PO Comparison

In this section, the various retransmission strategies, namely S1, S2 and MU, are compared when all the users have the same SNR.

Figure 5.7 shows that SU strategy (as in [23]) has a better performance in the low SNR region. S1, S2 and MU strategies outperform the SU strategy in the medium to high SNR range where the SU strategy lacks to exploit the multi-user diversity.

SU is the best choice in the low SNR region because retransmitting to the same user implies an higher probability of packets in error and thus higher probability of using double diversity transmission (STBC). When STBC is used, the errors are quickly recovered and the system can send two new packets. When the STBC is less used and its contribution is vanishing

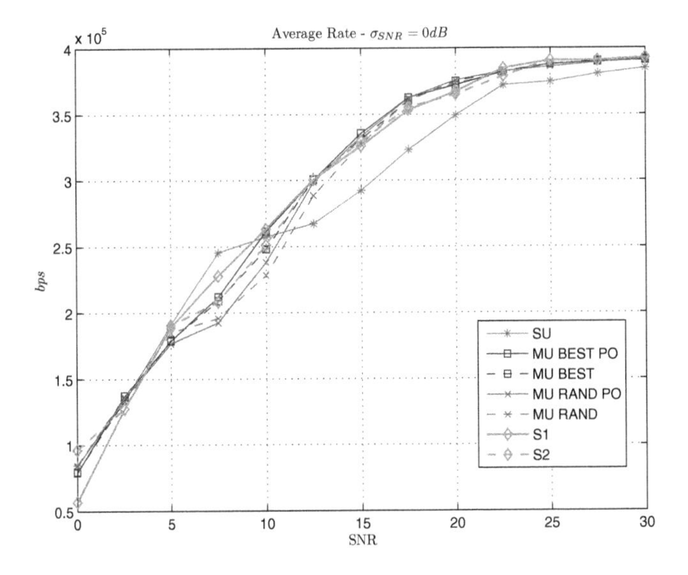

Figure 5.7 Average rate for S1, S2 and MU retransmission strategies

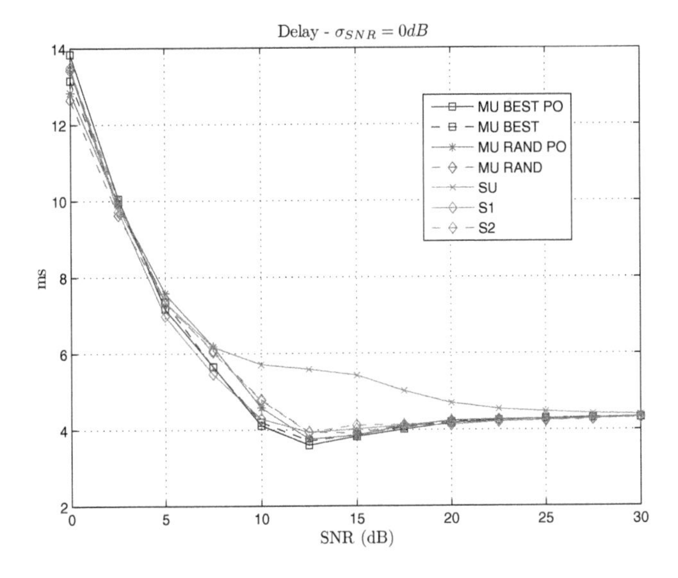

Figure 5.8 Average delay for S1, S2 and MU retransmission strategies

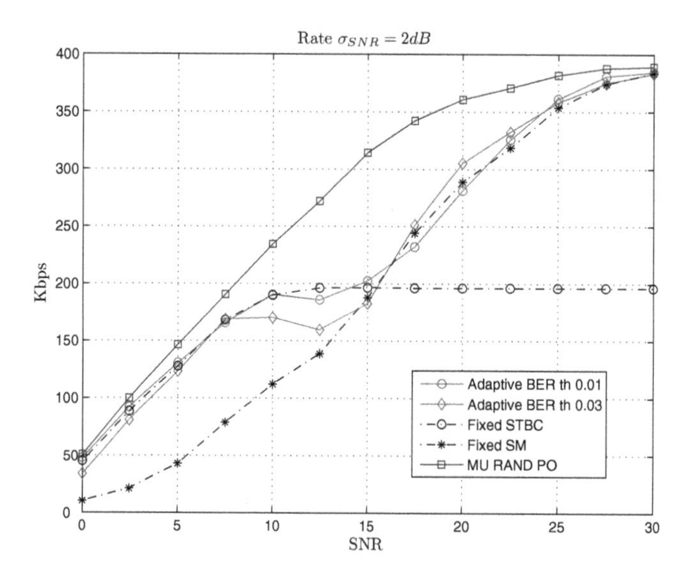

Figure 5.9 Average rate for MU MIMO-ARQ compared with non-adaptive and BER-adaptive systems with 2 dB log-normal SNR distribution

(SNR > 10 dB), SU performance quickly drops and S1, S2 and MU strategies show around 3 dB gain.[2] SU strategy also shows an higher average delay as reported in Figure 5.8.

The three strategies (S1, S2, MU) and the PO and non-PO receivers have comparable performance when all the user have the same SNR, with slightly better results for the MU strategy. When the user have different SNRs, as reported in Section 5.7.3, S1 and S2 will show sensible performance loss.

5.7.3 Performance with Log-Normal SNR Distribution

The MIMO-ARQ protocol proposed is able to cope with heterogeneous SNR distribution of the users in the system. In Figure 5.9 the average rate of the MU MIMO-ARQ with random user selection is compared with fixed and adaptive systems when the SNR follows a log-normal distribution with a standard deviation of $\sigma_{\text{SNR}} = 2$ dB. The adaptive system chooses the MIMO scheme based on the average SNR calculated over all the users. In Figure 5.10

[2] The system studied in this work is 4-QAM uncoded; at 10 dB is BER ≈ 0.03. A proper channel coding can move the cross-over point between SU and MU strategies at lower SNRs comparable with the typical operating point of the wireless systems.

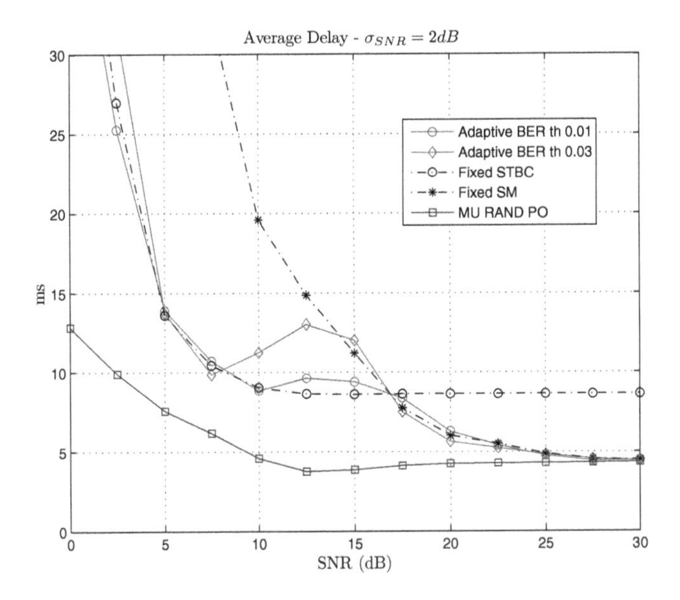

Figure 5.10 Average delay for MU MIMO-ARQ compared with non-adaptive and BER-adaptive systems with 2 dB log-normal SNR distribution

a tangible delay reduction can be observed compared to the adaptive systems. This comes from the hybrid SM transmissions the and exploitation of MU diversity which has a bigger impact compared to the case of $\sigma_{\text{SNR}} = 0$ dB. When $5 \leq \text{SNR} \leq 10$ dB, a 5 ms improvement is available, which means that in average users are served 10 packets earlier and the 4 ms average delay is already reached at SNR $\approx$ 10 dB. The delay jitter is also improved, as shown in Figure 5.11. Fixed STBC has, the lowest jitter but the jitter of MU strategy is still low compared to the fixed SM and the adaptive systems.

Figure 5.12 reports the average rate for all the strategies when $\sigma_{\text{SNR}} =$ 2 dB. MU strategy shows the best performance while S1 and S2 show a considerable reduction of performance of around 4 dB, starting from 12 dB. The MU with PO and BEST user selection also achieves slightly better rate compared to the RAND user selection due to optimal user selection for throughput maximization.

5.7.4 Performance for Short Channel Coherence Times

The MU MIMO-ARQ protocol has been evaluated also for short channel coherence times. Two coherence periods of 250 packets ($\approx$ 100 ms) and of

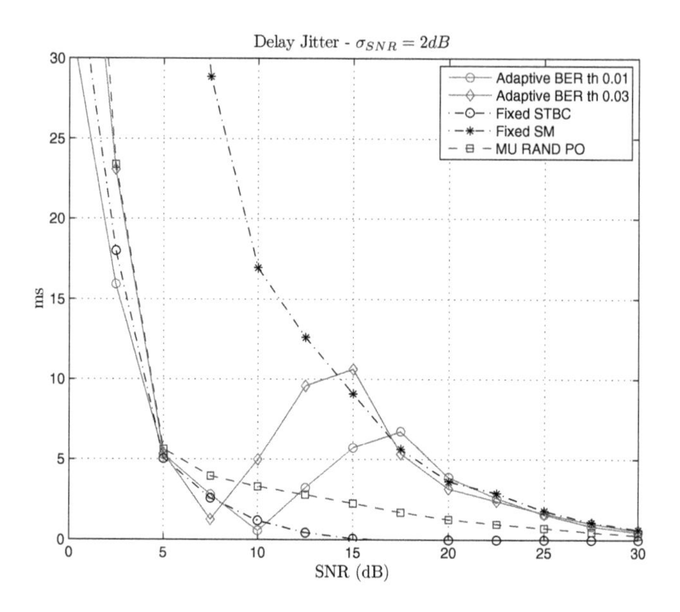

Figure 5.11 Delay jitter for MU MIMO-ARQ compared with non-adaptive and BER-adaptive systems with 2 dB log-normal SNR distribution

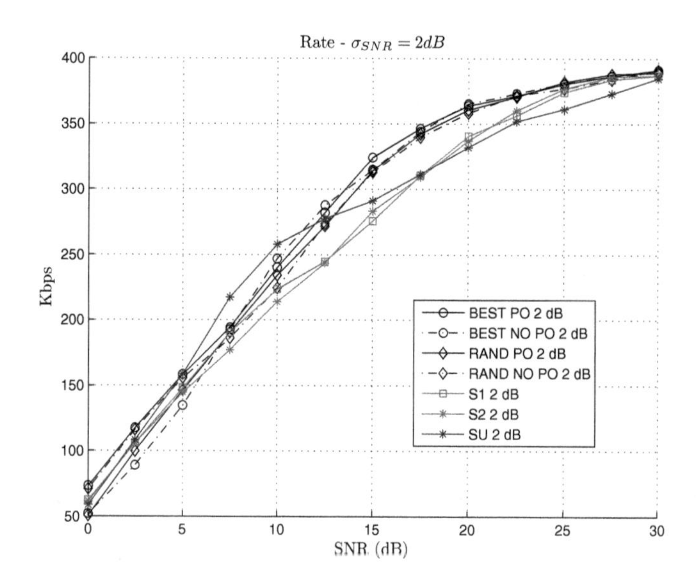

Figure 5.12 Average rate for S1, S2 and MU retransmission strategies with 2 dB log-normal SNR distribution

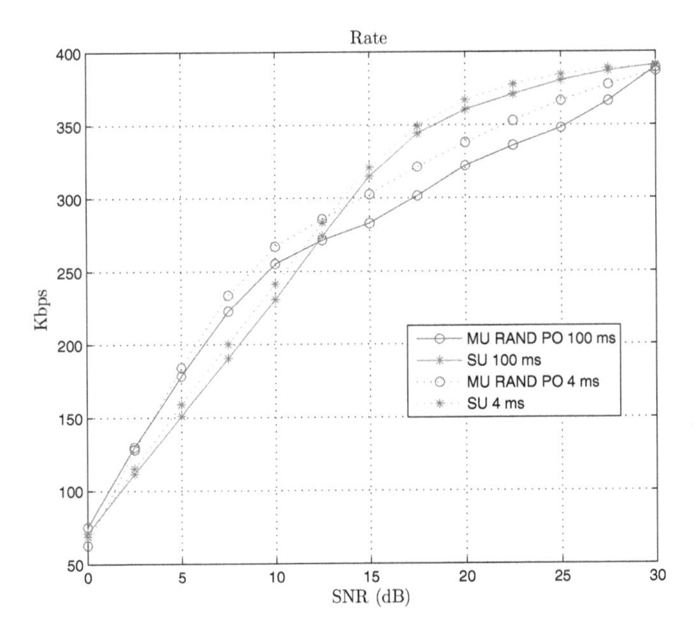

Figure 5.13 Average rate for MIMO-ARQ in MU and SU modes for short channel coherence times

5 packets ($\approx$ 4 ms) are compared. The performance of MU strategy with random user selection and SU (as in [23]) are compared. The results is that MU strategy is more robust the fast changing channels. The average rate is reported in Figure 5.13, where at 15 dB for 4 ms coherence time, the SU shows a loss of 2.5 dB against the 0.5 dB of the MU strategy. The reduction of average delay is maintain also with log-normal shadowing, as visible in Figure 5.14.

5.8 Conclusion

A MIMO-ARQ protocol has been developed for a MU network with three retransmission strategies. An optimized receiver, able to store previous transmitted packets, has been developed (Packet Overhearing). The users are scheduled with a RR scheduler; the system proposed can be directly extended including any kind of known scheduler.

The work extends the results achieved in [23], where retransmission strategy was designed for a SU link. The antenna allocation criterion was ex-

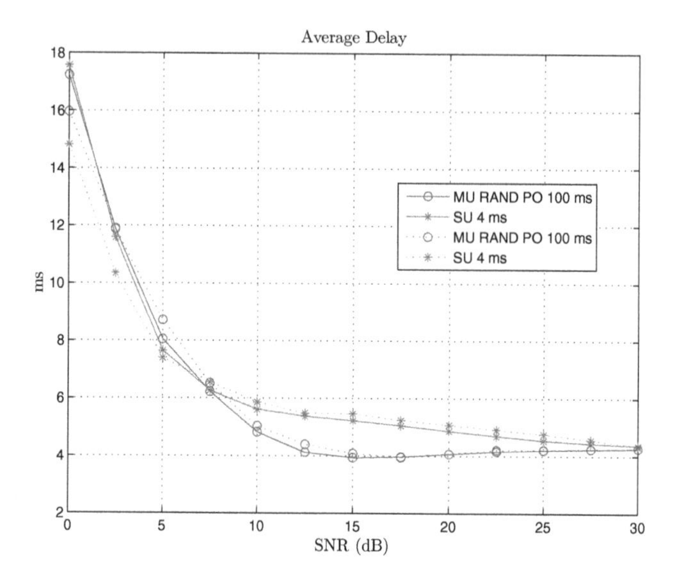

Figure 5.14 Average Delay for MIMO-ARQ in MU and SU modes for short channel coherence times

tended for MU networks. The MU MIMO-ARQ protocol can operate without any feedback (with the exception of the ACKs/NACKs) and avoids the one-bit feedback needed in [23]. Three strategies were proposed: retransmission from single best antenna (which still needs one bit-feedback), retransmission from boths antenna in spatial repetition coding and MU strategy, which uses hybrid SM transmissions.

In the last strategy, MU diversity is exploited with a random (RAND) user selection and no feedback is needed at the BS. The performance is compared with the BEST user selection for throughput maximization based on post-processing SNR measured at each MT. The performance degradation of the random user selection is small compared to the best user selection.

The MIMO-ARQ strategies for SU and MU are compared to wireless systems where fixed STBC/SM transmissions are used and to an adaptive system where the former two schemes are adaptively chosen to maintain the BER under a specified target. Better throughput and lower average delay are achieved over all the SNR range with a peak gain of 7 dB.

The protocol proposed outperforms the one in [23] for medium to high SNRs. At low SNR, strategy in [23] achieves better throughput. A cross-over point between MU and SU strategies is found at 10 dB for a 4-QAM uncoded modulation.

MU strategy outperforms the other strategy proposed (S1, S2), especially in the case of users with heterogeneous SNR. In log-normal SNR distribution, the MU gain over the SU strategy is more tangible due to the increasing impact of multi-user diversity exploited via random user selection.

The protocol proposed has been designed to maintain a good performance with fast time-varying channels (open-loop MIMO schemes are seamlessly used with no CSI feedback). The MIMO-ARQ protocol proposed can adapt rapidly to the channel status with no need of CSI feedback while exploiting multi-user diversity.

6

MIMO Protocol Gain System Design Issues

The adoption of MIMO-ARQ CL strategies presented in the previous chapter requires a reconsideration of the present layered network stack. This chapter will present the key ideas from an architectural point of view that will enable the gain offered by multi-user diversity exploitation inside adaptive MIMO coding schemes. The first communication network that was designed during the primitive years of information theory was the telephone network composed of a set of point-to-point links with fixed allocated resource. In such a system, when a call is originated, a path is reserved and the communication happens with no restrictions inside the assigned bandwidth.

The origin of packet-switched networks and the concept of packet payload itself was born because of the need for interconnecting computing machines. The burstiness of voice calls traveling over analog connection was, by far, much more reduced than that of the messages originated by the machines. Kleinrock was among the first to address this problem, formulating the concept of the backbone elements of a packet-switched network, which led to the well-known queueing theory, later published in [58].

During the late 1960s and early 1970s, several computer scientists and engineers proposed the formulation of the seven-layer Open System Interconnection (OSI) framework. At that time, the separation between the problem of physical data transmission and link management was very useful. The idea of layers is even found in the work by Shannon [96]. Shannon clearly stated that the digital channel that he was studying was a layer above the underlying physical channel where the data was transmitted in analog form. Shannon also considered the process of channel coding as residing a layer below the process of data compression.

However, for a number of years, there has been a strong revisionist feeling with respect to the notion of layering. The research community has fully realized that the strict layering of a packet-switched network leads to under-exploitation of system resource. Nonetheless, the layered framework offers a stable basis for technology development and standardization.

The constraints introduced by the protocol stack are clear in some cases and less evident in others. It is evident that the OSI layering was not designed for multi-access mediums, such as the wireless mediums. The optimal exploitation of multi-access networks requires a joint design of physical, link and network mechanisms. When multiple antenna or sectorized cells are available, the spatial dimension of the channel can be exploited and SDMA can offer high gains when designed with a CL approach.

The field of multi-user detection theory represents the link between multi-access information theory and network multi-access. There has been considerable activity on signal modelling, power-control, antenna patterns, channel interference models, and receiver structures.

The provisioning of a channel for an end-to-end transmission can be divided in several subtasks; traditionally each one of this tasks is accomplished in a separated layer of the stack. The PHY layer provides a bit tunnel for transmission adapted to the physical medium used. The organization of bits in packets is in charge of the Data Link Control (DLC) layer which provides reliable transmission service to the NET layer. The route from a source to a destination is provided by the NET layer, it manages a set of ordered links, each one with its own DLC and PHY processing.

The OSI layering was designed for wired-networks, and the stack needed to be modified to add the MAC layer to support communication over multi-access channel, e.g. as in the case of the ALOHA protocol in [3] just to cite a famous example. The MAC resides between the DLC and the PHY; basically, it provides an intermittent data connection to the DLC layer.

The MAC layer can be totally unaware of the underlying PHY or some information can be shared. This is the example of CQI signalling between PHY and MAC for channel-aware schedulers. Other example is the joint design of ARQ parameters based on CSI, as described in Section 4.2.

An interaction among PHY and MAC layers has been proposed in Chapter 5 when multiple antenna are available. In particular, the MAC is aware of a per-antenna ACK and several retransmission strategies have been designed for aggressive resource exploitation in multi-user networks with fast mobile terminals without the need for feedback.

The control of the ARQ process at the antenna elements level has some impact on the network architecture which will not require substantial modification to the current approach. In the current PHY/MAC layer division, it is not possible to be aware of the per-antenna channel conditions at the MAC layer. The designs described in Chapter 5 are analyzed from the signaling and architectural perspective.

The description of the current architecture refers to the last approved version of the IEEE802.16 standard [2]. Two topologies are defined for 802.16 networks: Mesh and PMP. Only the latter is considered in this chapter. The PMP traffic management, basic packet processing operations and the ARQ process are described. The ARQ state machine is controlled by the ACKs sent back by the MTs. ACKs (NACKs) are calculated based on the CRC appended to the MAC PDUs, as visible in the diagram in Figure 6.1. A modified interface between PHY and MAC is presented after (see Figure 6.11), in order to support the SU and MU MIMO-ARQ protocols of Chapter 5.

An overview of classic schedulers for IEEE802.16 networks is also proposed. This will clarify the possibilities offered by the integration of MU MIMO-ARQ system with more advanced schedulers compared to the RR scheduling assumed in the previous chapter.

6.1 General Overview of IEEE802.16 Point-to-MultiPoint

In IEEE802.16, the BS is usually a central point with a sectorized antenna system capable of handling multiple independent sectors simultaneously. Inside a given frequency channel and antenna sector, all MTs receive the same transmissions in DL. In UL, the BS collects all the signals coming from the MTs previously associated and which are inside the same sector. Only one BS is operating in the sector and does not have to coordinate with other transmitting stations. For DL, the DL-MAP specifies the addressee (MT) for each portion of the time-frequency resources. In the case where the DL-MAP does not explicitly indicate that a portion of the DL subframe is for a specific MT, all MTs will listen. The MTs decode the connection identifiers (CIDs) in the received MAC PDUs and store only those PDUs addressed to them. Messages can be sent to a single MT (unicast) or to multiple MTs (multi-cast connections). The latter is used for control messages and broadcast messages to all stations in the sector/cell.

In the UL subframe, all the MTs share the UL channels on a demand basis. The right to transmit depends on the class of service flow which will access the transmission bandwidth: the MT may be issued continuing rights to transmit, or the right to transmit may be granted by the BS after receipt of a request from the user.

MTs adhere to a transmission protocol that controls contentions and enables the service scheduling to be adjusted to the QoS requirements in terms of delay and bandwidth. There are three different types of UL scheduling mechanisms:

- Unsolicited bandwidth grants
- Periodic polling
- Contention based.

These three UL scheduling mechanisms allow vendors to optimize the system performance by using combinations of all the scheduling techniques, maintaining interoperability. For Unsolicited grants, the service flows has a fixed amount of bandwidth reserved independently from the load of the network and the status of the data queues. The use of polling guarantees that service flows receive resources on a deterministic basis when required. Contrary to data applications, which are delay tolerant, real-time applications, like voice and video, require service with very tightly-controlled schedule for preserving delay-related QoS (average delay and jitters) and need to have a constant check of the bandwidth requests (periodic polling). Finally, contention is used to avoid individual polling of MTs that have been inactive for a long period of time. The MT can send a request to the BS for bandwidth allocation in contention with the others MTs in the sector/cell in a specified part of the UL frame.

6.1.1 Packet Processing Workflow

The workflow for the packets to be transmitted over the wireless segment follows the scheme depicted in Figure 6.1, where a typical IEEE802.16 protocol stack is reported with the ARQ state machine in evidence. From the Application layer, packets are received at the air-interface. Following the IEEE802 nomenclature, each layer receives packets to be dispatched from upper layer (SDUs) and transfer to the lower layer fixed size packets to be transmitted (PDUs).

From the Networking layer, the NET PDU are passed to the MAC layer. The MAC layer is divided into three sublayers. The MAC Convergence Sublayer divides the traffic based on QoS attributes and provide transparent access to the wireless transmission from the transport perspective.

The MAC privacy sublayer provides support for identification and association to the network and for enforcement of security protocols. The security aspect is not treated here.

The MAC Sublayer is the central entity which covers the aspects related to the medium access. At this sublayer, NET PDU are packed or fragmented (de-

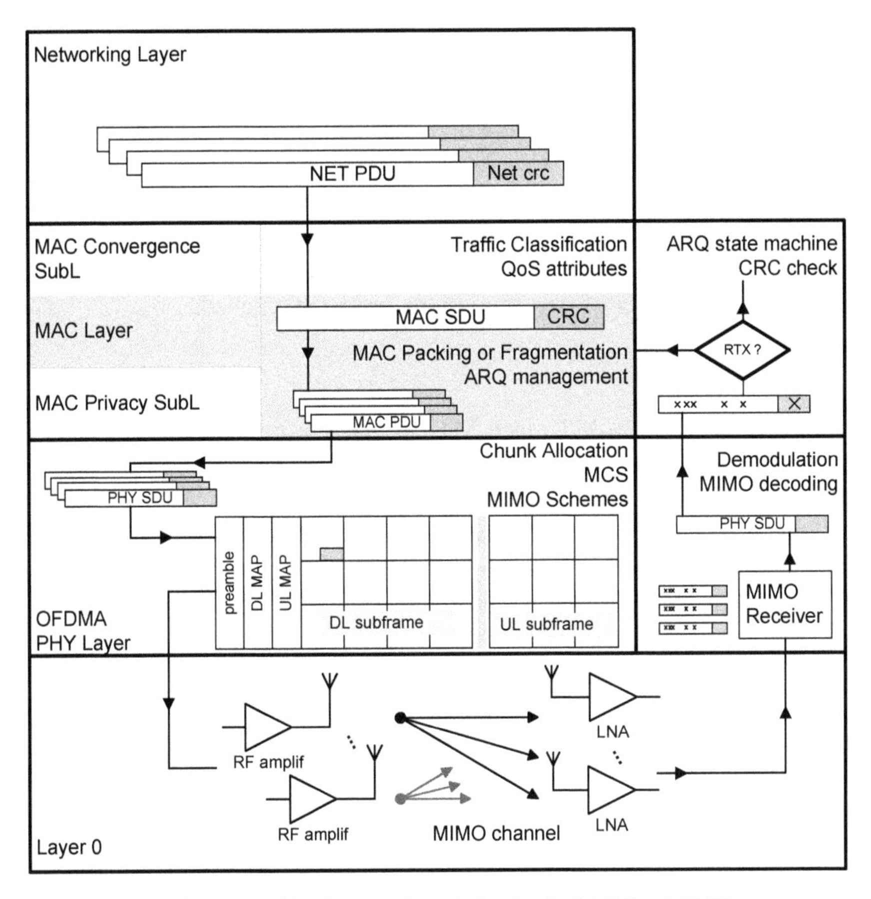

Figure 6.1 IEEE802.16 protocol stack for classical ARQ at MAC layer

pending on the size of the packets to be transmitted) in MAC PDUs which are later transferred to the PHY layer. The ARQ process is managed at this level of the protocol stack. Each MAC SDU, after being fragmented or packed, is appended with a CRC which is used to detect error during transmission on the radio channel. Fragmented/packed MAC SDUs are passed to the PHY layer. Each packet is then mapped in the OFDMA grid. Data randomization, modulation, subcarrier permutation and MIMO scheme are applied. Prior to be sent at the RF transmitting chains, the frame is composed along with data symbols, pilots and guard bands. The DL and UL subframes are separated by appropriate timings for permitting a correct ramp-up and ramp-down of the power amplifiers. The frame is left-padded with a Frame Preamble which is used for synchronization and coarse channel estimation purposes.

On the receiver side, the signal is processed with the appropriate MIMO receiver. Demodulation, decoding and depermutation is performed. A single PHY SDU is passed to the upper MAC layer for the CRC check. The result of this check (ACK/NACK) is then sent back to the BS. The ARQ state machine is fully contained at the MAC layer of the BS. If a NACK is received, a retransmission of packet occurs until the packet is correctly received or until the maximum number of retransmissions has been reached.

6.1.2 MAC Layer

Each air-interface of a MT (each MT can have more than one air-interface) has a 48-bit universal MAC address, as defined in the IEEE802 standard [1]. The MAC address uniquely defines the air interface of the MT in the network and it is issued by the manufacturer. The MAC address is used for the initial ranging and association processes and to establish the connections between the BS and the MT. During the authentication process, the MAC address is also used to verify the identity of BS and MT.

IEEE802.16 MAC is connection-oriented: all data communications are in the context of a transport identified with a CID. Upon the registration of an MT (or the modification of the services already negotiated), higher layers of the BS and of the MT are stimulated to initiate the setup of the service flows. Each service flow is associated with a transport connection and each MT can have more than one connection; each service flow is identified by a unique 16-bit CID. Bandwidth requests are always referred to a single CID. New transport connections are established for new traffic flows or when a MT changes the QoS attributes of an already established service flow. Indeed, the service flow contains all the QoS attributes for the PDUs that are exchanged on the MAC connection. Bandwidth is always granted by the BS to an MT (also as an aggregate of grants) in response to per-connection requests from the MT.

In the initialization phase, two pairs of management connections, basic connections (UL and DL) and primary management connections (UL and DL), are established between the MT and the BS.[1] The management connection pairs are used for three different levels of QoS of the management traffic. The basic connection is used to exchange short and urgent

[1] Optionally, a third pair of management connections (secondary management, DL and UL) can be generated.

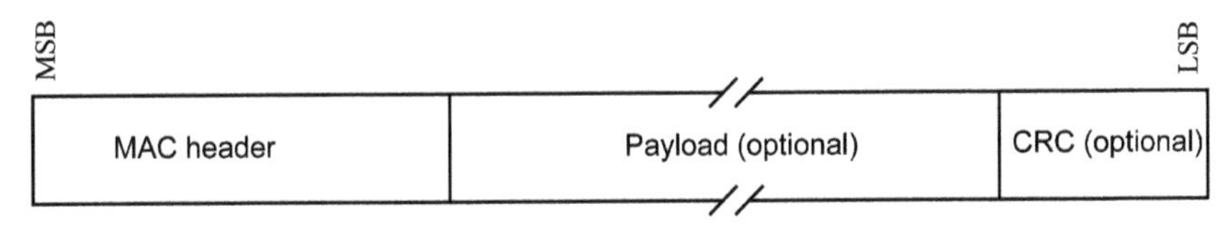

Figure 6.2 MAC PDU content in 802.16e

MAC management messages. The primary management connection is used to transport longer but more delay-tolerant MAC management messages. Table 38 in [2] specifies on which connection pair MAC messages have to be transferred, based on messages length and urgency. The table also specifies which MAC management messages are transported on the broadcast connection. An optional secondary management connection is used to transfer delay-tolerant messages of specific standards, such as: Dynamic Host Control Protocol (DHCP), File Transfer Protocol (FTP) (only trivial messages) and Simple Network Management Protocol (SNMP). The secondary management connection allows for packet packing or fragmentation. When OFDM and OFDMA physical layers are used, all the management messages must have an appended CRC.

Once established, transport connections for variable-rate data flows may require maintenance. For example, in the case of an unchannelized T1[2] service, virtually no connection maintenance is needed, since a constant bandwidth is allocated. IP services usually require a considerable amount of maintenance due to their bursty nature and the possibility of fragmentation. Transport connections may also be terminated: the termination of a transport connection is stimulated by the BS or by the MT. The transport connection management functions are implemented through the use of dynamic addition, deletion and modification of service flows, as described in Section 4.3.

The format of a MAC PDU is reported in Figure 6.2. The first part of the PDU is the General Mac Header (GMH) which is exploded in detail in Figure 6.3. The GMH contains several data fields, e.g., the PDU length in bytes, the CID and the Header Check Sequence (HCS) for detecting error in the GMH. For a detailed description of all the fields of the GMH, the reader is referred to [2, section 6.3.2.1.1]. In the case of fragmentation, each fragment of the SDU is encapsulated into one or more PDUs, as illustrated

[2] The historical definition of T1 service is related to a digital transmission line with 1.544 Mbps data rate equivalent to 24 voice channels used in telephone network; an unchannelized T1 service defines a service flow with 1.5 Mbps bandwidth.

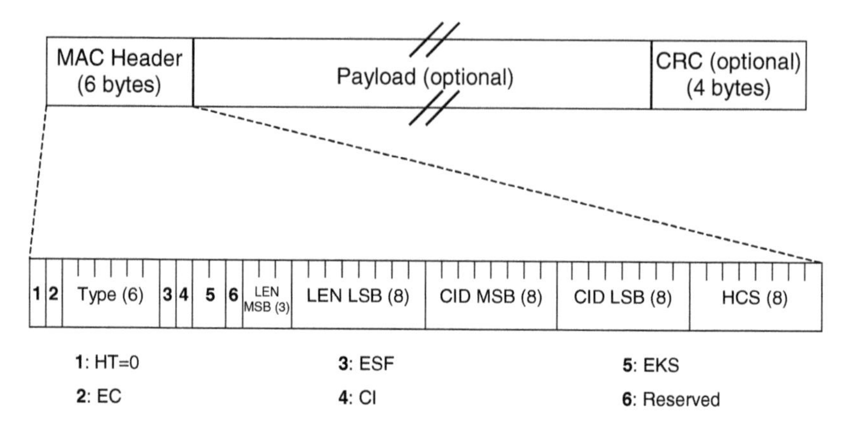

Figure 6.3 Generic MAC Header Content.

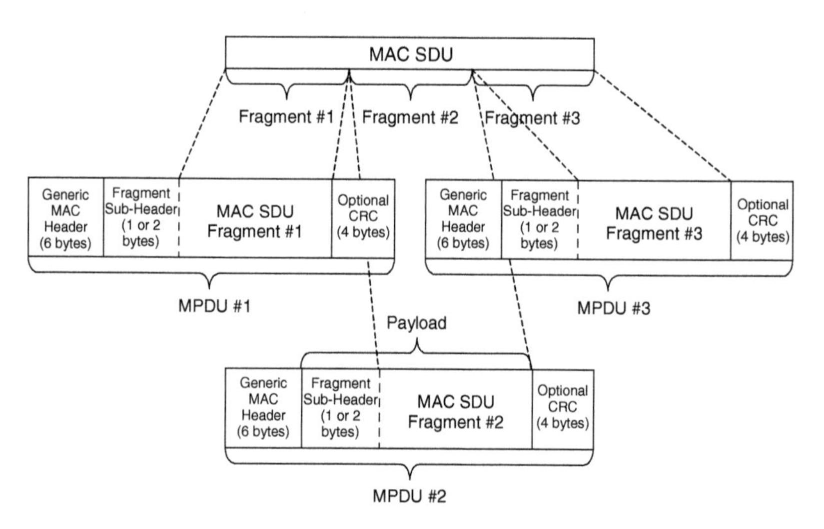

Figure 6.4 Fragmentation procedure at the MAC layer. Reproduced from [85]

in Figure 6.4. If SDUs are very small, more than one of them can also be encapsulated into a single PDU (packing), as reported in Figure 6.5. Internal Packing Sub-Headers (PSHs) are inserted.

Generic MAC Header (6 bytes)	Packing Sub-Header (2 or 3 bytes)	MAC SDU	Packing Sub-Header (2 or 3 bytes)	MAC SDU	....	Optional CRC (4 bytes)

Figure 6.5 Packed SDU into a single PDU

6.1.3 ARQ

The ARQ mechanism is part of the MAC layer. ARQ is optional for implementation. If implemented, ARQ may be enabled on a each single transport connection. The per-connection ARQ is specified and negotiated during the creation of the connection and is limited to unidirectional traffic. A single connection cannot have ARQ and non-ARQ traffic mixed together.

For ARQ-enabled connections, is possible to activate fragmentation of MAC SDUs: each SDU can be split into fragments for separate transmission. If fragmentation is not enabled, each PDU must contain all the data related to the parent SDU. The size of each MAC PDU is defined by the *ARQ_BLOCK_SIZE* parameter. An ARQ block is a distinct unit of data that is carried on an ARQ-enabled connection. The block size is a parameter negotiated during connection establishment. If the length of the SDU is not an integer multiple of the *ARQ_BLOCK_SIZE* parameter, the last fragment is padded with zero payload until the correct length has been achieved. The ARQ block size needs to be accurately set to avoid losses of network efficiency. Each PDU may contain ARQ blocks that are transmitted for the first time as well as those being retransmitted. Every ARQ block has an assigned Block Sequence Number (BSN) and is managed as a distinct entity by the ARQ state machines.

A diagram of the ARQ state machine, which is located at the BS, is reported in Figure 6.6. Once a MAC SDU has been scheduled for transmission and MAC PDUs have been generated after eventual fragmentation/packing, the PDUs are sent to the PHY layer and the state machine moves to the "Outstanding" state, where it waits for the first response (NACK or ACK). If an ACK is received, the state machine is reset. If NACK is received, the state machine moves to the state "Retransmission". The PDU is then transmitted again and the machine waits again in "Outstanding" for a NACK or an ACK from the MT.

When a NACK is receiver or a maximum Waiting-Time is elapsed the state machine retransmit the packet. The maximum Waiting-Time is spe-

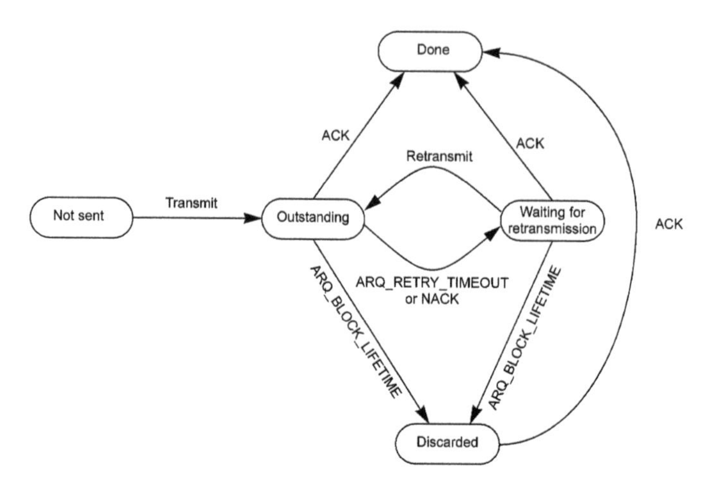

Figure 6.6 ARQ TX state machine. Reproduced from [2]

cified by *ARQ_RETRY_TIMEOUT*.[3] This parameter is fundamental to avoid starving of the state machine for two cases: (1) the PDU has not been received at the MT, (2) the packet was received at the MT and the MT transmitted the NACK/ACK which was not received at the BS. Finally, from the two intermediate states presented before, the state machine can interrupt the protocol operation when an ACK is received or when a maximum amount of time is elapsed for the ARQ process, specified by *ARQ_BLOCK_LIFETIME*. This specification is equivalent to specifying the maximum number of ARQ retransmission, as done for Stop and Wait ARQ in Section 4.2 and for MU MIMO-ARQ in Section 5.4. It is to be noted that an ACK signal also link the "Discarded" to the "Done" state. This is to include the case where a positive ACK is received slightly after the *ARQ_BLOCK_LIFETIME* was over.

A system supporting ARQ must be able to receive and process the ARQ feedback messages. The ARQ feedback information is sent back to the BS and cannot be fragmented. ARQ feedbacks can be implemented in two forms:

- on standalone MAC management connection (in UL)
- via piggyback on an existing UL data connection.

[3] The definition of *ARQ_RETRY_TIMEOUT* is specified in [2, section 6.3.4.3].

The ARQ feedbacks always refer to the ARQ BSN. Several ARQ feedback types are possible:

- selective
- cumulative
- cumulative and selective
- cumulative and sequence.

In the selective feedback, the MT acknowledges the ARQ blocks received with a single BSN value and four 16-bit selective ACK maps. The BSN value refers to the first block in the first map. The MT sets the bit of the selective ACK map to one or zero according to the reception of blocks without or with errors, respectively. The cumulative type can acknowledge any number of the ARQ blocks. The MT acknowledges a BSN for which all the packets with a lower or equal BSN have been correctly received. The "cumulative and selective" ARQ feedback just combines the functionality of the former two.

The last type, "cumulative and sequence", combines the functionality of the cumulative type with the ability to acknowledge reception of ARQ blocks in the form of block sequences. The ARQ block sequence is defined as the set of ARQ blocks with consecutive BSN values and which have been correctly or incorrectly received by the MT. The sequence ACK map entity contains a bit that indicates that a corresponding block sequence has been received without errors (or with errors) and the sequence length which reports the number of block that are members of the ARQ block sequence. The choice among ARQ feedback types is left to the feedback sender. All the ARQ feedback types are mandatory with the exception of "selective ARQ".

Each ARQ feedback type has its drawbacks and advantages which depend on the feedback transmission frequency, the error distributions and the computational complexity that can be dedicated by the feedback sender – which is usually quite limited at the MT. The selective feedback type is the simplest and does not require a lot of processing resources. On the other hand, selective ARQ feedback may have low protocol efficiency. When resource are scarce, the "cumulative and sequence" feedback type should be used. Of course, the implementation is more complex because ARQ block sequences must be constructed analyzing the pattern of the ACKs and NACKs to be sent back.

An advanced features of the ARQ at the MAC layer, is the rearrangement of MAC PDUs, as shown in Figure 6.8.

An SDU is partitioned into a set of blocks prior entering the ARQ process and the partitioning remains in effect until the SDU is discarded by the

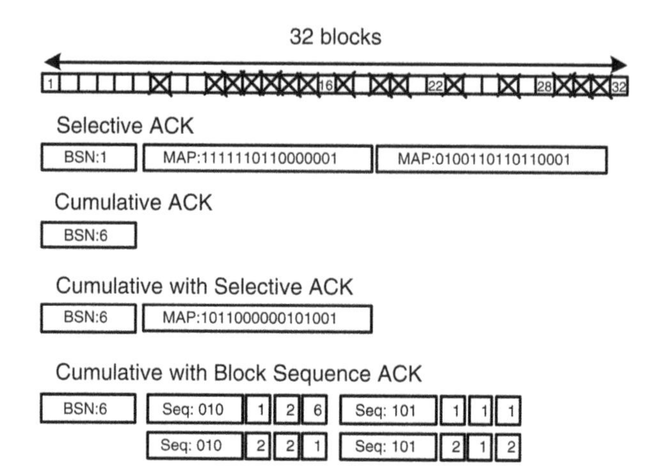

Figure 6.7 ARQ types in IEEE802.16 2009. Reproduced from [95]

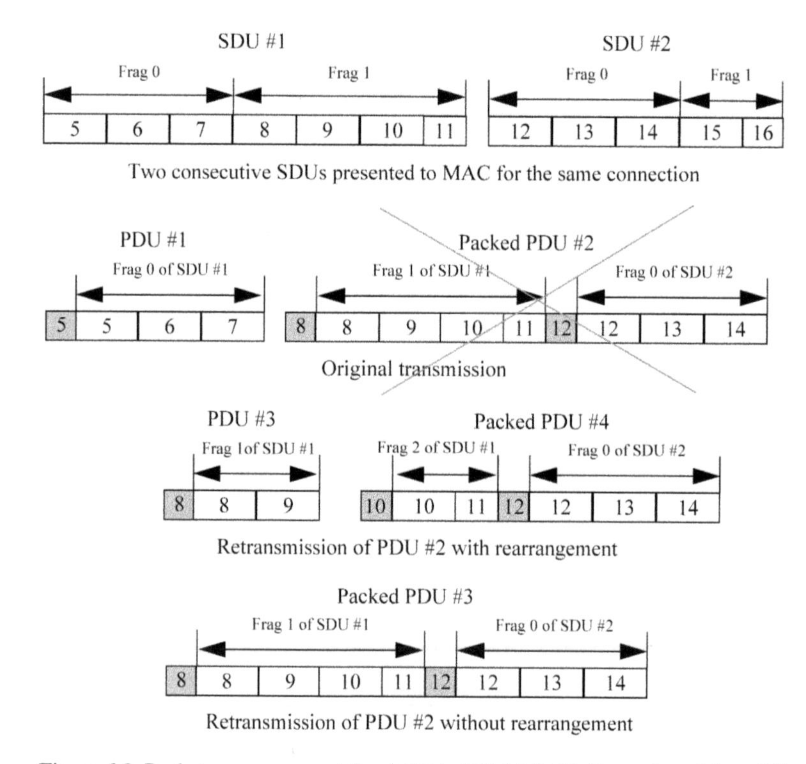

Figure 6.8 Packet rearrangement for ARQ in 802.16 2009. Reproduced from [2]

transmitter state machine (if maximum retransmission time is elapsed) or if all blocks of the SDU are successfully delivered to the receiver.

Part of ARQ blocks, selected for transmission or retransmission, are encapsulated into a PDU. When a PDU is composed, the sequence of blocks immediately between MAC subheaders must have contiguous block sequence numbers. However, if ARQ arrangement is available, the fragments of MAC SDUs can be rearranged on MAC PDUs in order to save bandwidth during retransmissions. This requires added processing at the MAC layer for both ends of the connection.

6.2 Scheduling in IEEE802.16

The scheduling service represents the mechanisms for data packets handling on a single connection. Each connection can be associated only with a single scheduling service. QoS parameters define a scheduling service (or service class). IEEE802.16 wireless networks defined several QoS classes. These classes are used to guarantee the required rate and delay performance for heterogeneous applications, such as: data, voice and video traffic flows.

IEEE802.16 standard does not include a definition for the scheduler implementation and leaves freedom to the vendors for differentiation. Several schedulers have been proposed in the literature with channel-aware and channel-unaware techniques, which need to interwork with PHY layer to obtain CSI information and channel-unaware techniques. Each scheduler is characterized by the following tasks:

- guarantees of the QoS
- maximizes the throughput
- optimizes the resource utilization
- minimizes the average delays and delay jitters.

All these features are to be achieved with the smallest computational complexity. Also, scalability is an important characteristic of the scheduler, i.e. the scheduler has to be complexity-limited and viable when an high number of connections is to be managed. The task is far from being easy, since the radio channel has an high degree of variability (especially with fast moving MTs).

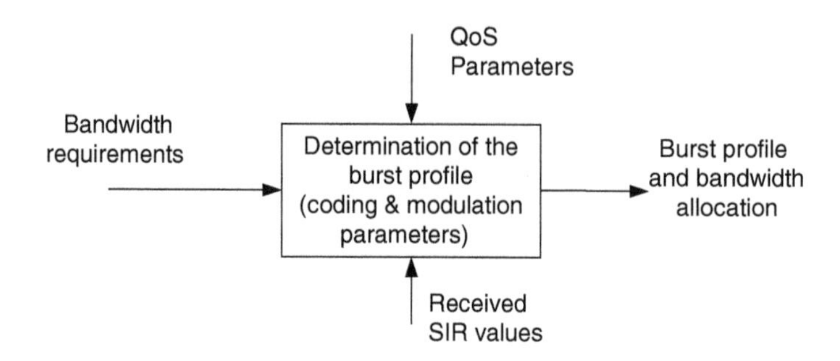

Figure 6.9 Input parameters for the scheduler

6.2.1 BS and MT Schedulers

In IEEE802.16 with PMP configuration, the BS is equipped with a sectorized antenna and can manage independently several sectors. Inside a single sector, the scheduler controls the admission of service flows and the MAC SDUs scheduling. It is also a role of the BS scheduler to control the adaptive modulation and coding of each flow and to enforce power control.

The MCS profile is decided by the BS based on several parameters:

- SIR
- Radio channel quality
- Network load
- Status of the packet queues.

The scheduler has to monitor constantly the received SIR values for each associated MT and determine the bandwidth to be allocated considering the service QoS attributes. Figure 6.9 reports a simple diagram summarizing the inputs and outputs parameters of the scheduler.

The BS scheduler decides for the DL and UL resource allocation. The DL allocation is broadcasted in the DL-MAP with MAC management messages. On the UL, based on the UL requests and taking into account QoS parameters and scheduling services priorities, the BS scheduler decides an UL resource allocation which is broadcasted in the UL-MAP. There is also a scheduler in the MT which have the role to classify all the incoming packets into the MT different UL connections.

The BS and MT schedulers are reported in the diagram of Figure 6.10. For each CID, a queue is set up in the scheduler memory space. At the BS, two schedulers manage the DL and UL connections separately. DL SDUs from the

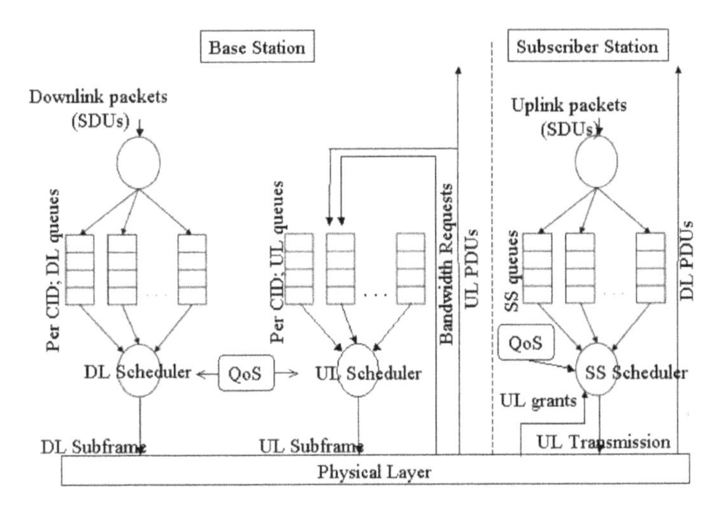

Figure 6.10 DL and UL schedulers in IEEE802.16, from [99].

upper layers are placed in the queues after eventual packing/fragmentation and CRC appending operations. Instead, received UL PDU are directly are transferred from the PHY to the MAC functions for de-fragmentation/de-packing, decoding and CRC checks. The UL scheduler at the BS is informed of the Bandwidth Request sent by the MTs in the previous UL subframes.

On the MT side, an UL scheduler is also present: UL PDUs generated at the MT are placed in the scheduler queues. After having received from the BS the UL bandwidth allocation grants in the previous DL subframes, the MT scheduler transfers the packets to be transmitted from the scheduler queues to the PHY layer. Received DL PDUs are directly transferred from the PHY to the MAC layer.

IEEE802.16 has not specified any packet scheduling or admission control mechanisms and several algorithms have been proposed in literature.

Schedulers for WiMAX can be classified into two main categories. The schedulers which do not have any information from the PHY layer regarding the CSI of the admitted users (channel-unaware schedulers) and the schedulers which accounts for the actual radio channel qualities (channel-aware schedulers). The channel-unaware schedulers do not use any information of the channel state condition in making the scheduling decision and generally assume error-free channel. In wireless networks there is a high variability of radio links (signal attenuation, noise, fading, interference) and the channel-awareness is important. An overview of the impact of channel awareness on the scheduling process can be found in [7].

Scheduling mechanisms have a big impact on network performance and a consistent amount of research is still ongoing; some recent works are herein cited. Several scheduling algorithms have been proposed such as Fair Scheduling [53, 66, 72, 103, 126], Distributed Fair Scheduling [17], MaxMin Fair Scheduling [107, 108], Customized Deficit Round Robin [62].

Another family of schedulers is based on traffic priorities calculated from the HOL delay and the averaged sustained data rate in respect to the QoS delay and rate requirements, as in [80, 118].

6.2.2 QoS Classes

The IEEE802.16-2009 standard MAC Layer is able to provide QoS differentiation for the various types of applications through five types of scheduling services, or QoS classes. This classification facilitates bandwidth sharing among different MTs. According to the QoS classes and the radio channel quality, the BS scheduler allocates the necessary amount of bandwidth required for each connection: e.g. a real-time application, such as a video or voice over IP application, will be prioritized compared to FTP or email or web traffic.[4]

In IEEE802.16d standard, four scheduling services were defined:

- Unsolicited Grant Service (UGS)
- real-time Polling Service (rtPS)
- non-real-time Polling Service (nrtPS)
- Best Effort (BE).

Each of these classes has a mandatory set of attributes (QoS parameters) that are included in the service flow definition. The mandatory service flow QoS parameters for each of the four scheduling services is defined in [47].

The UGS QoS class is designed to support real-time data with fixed-size data packets to be transmitted at periodic intervals, e.g. Voice over IP without silence suppression. In this case, the BS provides fixed-size data grants at periodic intervals. eliminating the overhead due to MT grant requests.

The nrtPS is designed for delay-tolerant data streams with variable-size data packets. For nrtPS a minimum data rate is defined. This is the situation,

[4] This feature is not included, for example, in Wi-Fi WLAN networks where all services have the same level of QoS.

Table 6.1 IEEE802.16 mandatory QoS calsses and attributes

QoS Classes	Maximum traffic rate	Minimum traffic rate	Tolerated jitter	Maximum delay	Traffic priority
BE	√				√
nrtPS	√	√			√
rtPS	√	√		√	
ertPS	√	√		√	
UGS	√		√	√	

for example, of an FTP connection. The BS polls nrtPS connections typically on an interval on the order of one second or less.

The rtPS scheduling class is used for delay sensitive service flows (i.e. video streaming) and it is designed to support real-time data streams with variable-sized data packets issued at periodic intervals. In this service, the BS provides periodic unicast (UL) request opportunities.

The BE service is transmitted on a best available basis. This class is designed to support data streams for which no minimum guarantees are required. When the network is congested, a long period can run without a BE connection receives any packet.

Another class, extended rtPS (ertPS), was added by the 802.16e amendment. The aim of ertPS is to reduce overhead for grant requests in rtPS: the BS provides unicast grants in an unsolicited manner like in UGS, thus saving the latency of a bandwidth request. Compared to UGS allocations, which have a fixed-allocation resource assigned, ertPS are dynamic providing variable bandwidths. The ertPS is suitable for variable rate real-time applications with rate and delay requirements, as VoIP with silence suppression. Table 6.1 reports the traffic attributes for each QoS service class.

6.3 MIMO-ARQ Protocol Stack

The MIMO-ARQ system proposed in Chapter 5 needs to be aware of a per-antenna ACK in order to implement the retransmissions strategies.

A modified stack is proposed here and illustrated in Figure 6.11. As every cross-layer technique, an enriched interface is needed between the PHY and MAC layer. The added signalling allows the ARQ state machine to be aware of MIMO transmission with a per-antenna ARQ process. Compared to the classical stack, as in Figure 6.1, more data pipes are available on the PHY/MAC boundaries, one for each antenna element. The resource allocation at a chunk level is still defined at the PHY layer, but the antenna allocation is informed by the MAC layer MIMO-ARQ state machine. The MIMO coding (STBC or SM) is decided at the MAC layer. The PHY layer places the multiple MAC PDUs received to the transmitting antenna as instructed by the MAC which follows one of the strategies presented in Chapter 5. The PHY layer modulates the payloads accordingly to the MIMO-ARQ transmission strategy.

At the receiver side, the MIMO receiver at the PHY layer processes the packets and sends several PHY SDUs to the upper layer for CRC checking (one for each receiving antenna). The CRC is checked for each packet and multiple ACKs/NACKs are sent back to the BS. The MIMO-ARQ state machine resides at the MAC level at the BS but it is aware of ACK/NACK for each antenna as sent back by the MT (or MTs in the case of hybrid SM transmission).

At the MT, a Packet Combining (PC) repository is available, where all the packets previously transmitted are stored; the packets correctly received are cancelled from the received signals, as explained in Section 5.5.5.

The MU MIMO-ARQ protocol proposed in the previous chapter is designed to be readily integrated with advanced packet schedulers. In particular, as shown in Figure 5.1, two levers are available.

The performance gains shown in the previous chapter were obtained considering a simple RR scheduler and random user selection for hybrid SM transmission and the BEST user selection was solely analyzed as a performance upper-bound of the previous un-informed user selection method.

Despite the perfect fairness provided, the RR scheduler lacks to exploit the network resource with heterogeneous traffic flows. A more advanced scheduler could enhance the average data rate via multi-user opportunistic scheduling.

An added opportunity is offered by the user selection method in hybrid SM transmission. Instead of being chosen randomly, the user could be chosen based on some traffic parameters measured at the packet queue. For instance, if a user is approaching its HOL maximum delay, the scheduler could promptly chose the user for the next hybrid SM transmission thus giving an opportunity of receiving a packet without a deep impact on the current transmissions ongoing.

Various possibilities are clearly available and more than one urgency level could be defined. A “medium” urgency could be treated with the scheduling for the next hybrid transmission while a “severe” urgency could be addressed scheduling the user directly in STBC, interrupting the current transmission and thus providing the highest reception probability.

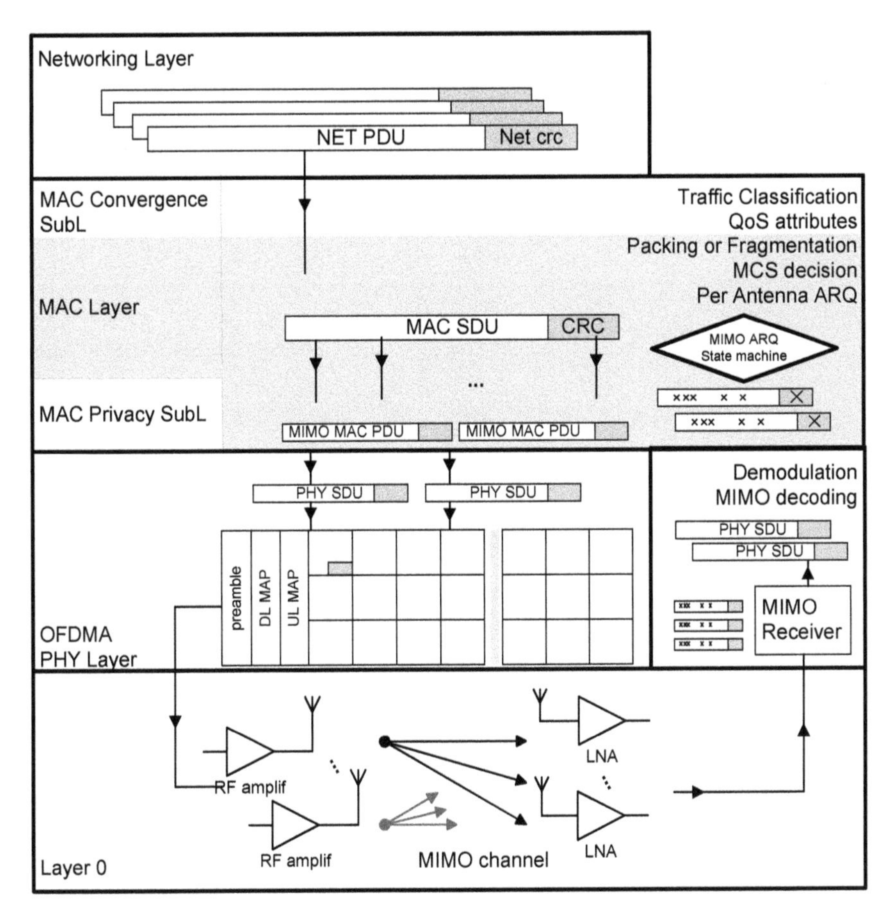

Figure 6.11 Modified protocol stack with per-antenna ARQ

The scheduler has an enlarged set of actions compared to a classical channel-aware scheduler. Indeed, the MIMO-ARQ scheduler can decide to:

- schedule the user in SM after the next MIMO-ARQ state machine termination
- schedule the user in STBC after the next MIMO-ARQ state machine termination
- schedule the user for the next hybrid SM transmission
- schedule the user in SM interrupting the current state machine evolution
- schedule the user in STBC interrupting the current state machine evolution.

The previous decision could be taken with all or some or none of the following: CSI, sustained data rate, average delay or average delay jitter measured at the MAC scheduler queue.

The added degree of complexity should be evaluated with memory and processing power constrains, particularly evaluating the scaling issues with the number of MTs admitted in the system.

7

Conclusions

The ongoing process of globalization requires interoperability of the wireless networks deployed around the world. The next-generation wireless systems will fulfill the specifications published by the International Telecommunication Union (ITU) defining the International Mobile Telecommunications Advanced (IMT-A) system that guarantees worldwide and intra-system interoperability.

IMT-A Wireless networks will be able to provide high Spectral Efficiency (SE) radio links toward Mobile Terminals (MTs) moving with velocities as high as 300 km/h. High transmission rates will be offered to stationary and mobile users, supporting the convergence of data, voice, and video transmissions to provide the "Mobile Internet", i.e., a connection over secure IP with the same performance of a cable-connected computer, but on the move. The convergence of heterogeneous services imposes strict requirements on the Quality of Service (QoS) that must be ensured to users' transmission. In wireless networks it is extremely difficult to guarantee such level of QoS requirements, especially due to the intermittent quality degradation that radio channel may experience, greatly exacerbated by the high user mobility.

In order to achieve the requested QoS, strategies that counteract the channel quality variations in an *adaptive* way are necessary. Such adaptive strategies can be designed at each layer of the stack separately. The choice of the layer at which the adaptive strategy is implemented affects the performance of the wireless network. In particular, strategies implemented at the upper layers – e.g., Internet Protocol (IP) layer – are too slow to counteract the channel variations of vehicular radio channels. In addition, if different disjoint strategies are implemented at different layers, lower layers – e.g., the physical (PHY) layer – can compensate the channel quality degradation very quickly and the upper layers may strive to achieve the QoS requirements and stabilize their adaptive algorithms.

The classic OSI protocol stack has been a fundamental basis for a coherent and stable technology development of practically viable communication systems. However, it is widely recognized that the strict division between physical (PHY) and Medium Access Control (MAC) layers is unsuitable for multiple access channels, which is the typical situation encountered in wireless cellular systems. The limitations are especially demanding when aggressive performance is to be achieved and multiple antennas are used. Indeed, Multiple Input Multiple Output (MIMO) enlarges the signaling dimensions and enriches the possibilities of access to the channel when the antenna elements are used for traffic directed to different users in Multiple User (MU) networks.

For the previous reasons, the adaptive strategies must feature a cross-layer approach to optimally achieve the QoS requirements and promptly compensate the channel quality variations. In an MU MIMO network, a cross-layer approach is advocated even more since it can greatly improve the performance exploiting the gain coming from Multi-User diversity.

In this work, a characterization of PHY signaling and MAC protocols has been proposed. A set of cross-layer designs have been presented which integrate the features of each layer and advance the state of the art of the strategies previously proposed for each layer.

In particular, a MIMO-ARQ protocol, with several retransmission strategies, has been proposed that integrates MIMO open-loop schemes jointly with ARQ process and packet scheduling. The MIMO-ARQ protocol showed increased network throughput, reduced average delay and reduced delay jitters over all the SNR range compared to fixed MIMO schemes and to adaptive MIMO schemes applied at the PHY layer, independently from upper layers.

Throughout the monograph multiple results were discussed which are summarized here. First, at the PHY layer, a dynamic MIMO scheme selection has been designed with an upper-bounded BER. The system switches among several PHY profiles using Space Time Block Code (STBC), Layered STBC (LSTBC) and Spatial Multiplexing (SM) schemes in order to provide the highest SE maintaining the BER under a specified target. When no multiplexing gains are required, the quasi-orthogonal schemes should be avoided in order to have better throughput. Data throughput and SM gain can be traded off based on channel quality. The performance of the adaptive PHY layer is able to boost the PHY layer performance but cannot address or mitigate the problem of low MIMO channel ranks due to low scattering. This adaptive strategy is later compared with the CL MIMO-ARQ protocol. The impact of a dynamic PHY layer is to be included in the design of the protocol stack since the MAC layer has to adapt promptly to the variations of SE at the MAC/PHY interface.

Then, dynamic allocation of Space Time Block Codes (STBCs) and Space-Frequency Block Codings (SFBCs) has been studied for an OFDMA PHY layer in order to preserve the performance of Space-Time Codes (STCs). Based on delay spread and Doppler frequency measurements, which can be easily obtained in OFDM Multi-Carrier systems as recently shown in literature, the diversity order of the STC scheme is preserved by allocating the symbols along the direction (time or frequency) that shows the highest correlation. The correlation coefficients have been derived under the Wide Sense Stationary – Uncorrelated Scattering (WSSUS) channel assumption. It is shown that at velocities over 120 km/h, comparable correlation is measured in the time and frequency domains. This is a fundamental indication for the design of high-velocity radio links with STCs that will adapt to the time-frequency characteristics of the MIMO channel. The proposed STBC/SFBC adaptive system needs a single-bit feedback. The ratio between the spatial coherence of the channel delay spread and the MT velocity gives the required feedback rate. Such lightweight feedback is an important design feature to implement this solution with fast MTs.

At the MAC, two protocols have been analyzed with the queueing theory: Automatic Repeat Request (ARQ) and Dynamic Service Addition (DSA). The number of ARQ retransmissions has been optimized with the selection of one of several Single Input Single Output (SISO) PHY profiles. The proposed Adaptive Cross Layer (ACL) strategy jointly and promptly optimizes the PHY and MAC parameters with the objective of guaranteeing the QoS requirements while allocating the minimum amount of bandwidth necessary for transmissions. Even without the use of MIMO, QoS requirements are met on a larger SNR range compared to the non-cross-layer adaptation, and the ACL is able to counterfight the channel shadowing very efficiently. Eight times less bandwidth is needed to transmit the same amount of data when the channel shows a log-normal shadowing with a standard deviation of 2 dB. A counterintuitive result has been observed: channels with high SNR deviation show lower average delays, and mobility can impact positively on the average transport delay if an ACL strategy is implemented with the ARQ protocol.

Again at the MAC layer, the DSA protocol have been analyzed. When harsh radio propagation is experienced, the DSA protocol can lead to high signaling blocking probabilities, thus preventing service flow from entering the network and reducing the resources utilization factor. A closed-form signaling blocking, admission control blocking and latency have been derived, also in the case of the activation of the service flow with an uninformed MT. The theoretical model has been fully validated for a block fading channel. The impact of mobility, network load, DL/UL reciprocity, OFDM PHY layer parameters – such as the DFT size – have been studied over fast channels, and the impact on the blocking probabilities and latency has been determined for slow and fast MTs.

Finally, an interaction among PHY and MAC layers has been designed to exploit Multi-User diversity with open-loop MIMO schemes and ARQ. In particular, the MAC is aware of a per-antenna ACK and several retransmission strategies have been proposed for aggressive network exploitation without the need of any CSI feedback. The feedback is avoided by design: (1) for the optimal antenna selection for ARQ retransmissions and (2) for the user selection in Hybrid SM transmission mode: the user is chosen randomly. The random user selection shows small performance loss compared to the case where the best user is selected for network throughput maximization.

For MIMO-ARQ protocol, both Single User (SU) and MU networks are analyzed. The MU MIMO-ARQ protocol has shown very good performance even with channel coherence times as low as $4ms$: much lower than the minimum delay requirement for the data plane specified for IMT-A system in the case of a fully unloaded network.

Further, the cross-layer MU MIMO-ARQ protocol had shown a throughput gain of 7 dB over the single-layer adaptive strategy based on upper-bounded BER implemented independently at the PHY layer. Compared to single-layer adaptive strategies, the average packet delay of the MU MIMO-ARQ protocol is noticeably lower over all the SNR range.

The strength of the CL approach and the gains attained with the MU MIMO-ARQ protocol have been shown and compared to other adaptive strategies previously studied. The results are supported by several theoretical analyses which have been conducted for the following:

- closed-form calculation of the average Symbol Error Probability (SEP) at the PHY layer
- ARQ and DSA signaling blocking probability
- Average Delay of the ARQ protocol
- DSA admission control blocking probability
- DSA service flow activation probability with uninformed Mobile Terminal (MT)
- DSA service flow activation latency.

The SEP calculations at the PHY layer were conducted using alternative forms of the error and squared error-functions plugged in the Moment Generating Function (MGF) of the SNR distribution at the output of the MIMO receivers.

The system design issues have been discussed inside the IEEE802.16 framework (WiMAX), referring to the latest version of the standard. However, the theoretical approach followed in the proposed designs and the several common features shared by WiMAX and LTE-A make the results achieved in this work valid and applicable in a very broad context and directly related to the technical development of the IMT-A future systems.

Bibliography

[1] Information processing systems – Local area networks – Part 2: logical link control. ISO Std 8802-2: 1998; IEEE Std 802.2-1998, December 1989.

[2] IEEE Standard for Local and Metropolitan Area Networks Part 16: Air Interface for Broadband Wireless Access Systems. IEEE Std 802.16-2009 (Revision of IEEE Std 802.16-2004), pages C1–2004, 2009.

[3] Norman Abramson. The Aloha system: Another alternative for computer communications. In *AFIPS'70 (Fall): Proceedings of the November 17–19, 1970, Fall Joint Computer Conference*, pages 281–285, ACM, New York, NY, USA, 1970.

[4] S. Al-Ghadhban, R.M. Buehrer, and B.D. Woerner. Outage capacity comparison of multi-layered STBC and V-BLAST systems. In *Proceedings of IEEE Vehicular Technology Conference, VTC Fall*, 2005.

[5] S.M. Alamouti. A simple transmit diversity technique for wireless communications. *IEEE Journal on Selected Areas in Communications*, 16(8):1451–1458, October 1998.

[6] A. Annamalai and Vijay K. Bhargava. Simple and efficient techniques to implement a self-reconfigurable arqsystem in a slowly varying mobile radio environment. *Wireless Personal Communications*, 13(1-2):97–117, 2000.

[7] Leonardo Badia, Andrea Baiocchi, Alfredo Todini, Simone Merlin, Silvano Pupolin, Andrea Zanella, and Michele Zorzi. On the impact of physical layer awareness on scheduling and resource allocation in broadband multicellular IEEE 802.16 systems [Radio resource management and protocol engineering for IEEE 802.16]. *IEEE Transactions on Wireless Communications*, 14(1):36–43, February 2007.

[8] Xiaofeng Bai, A. Shami, and Yinghua Ye. Robust QoS control for single carrier PMP mode IEEE 802.16 systems. *IEEE Transactions on Mobile Computing*, 7(4):416–429, April 2008.

[9] P. Bello. Characterization of randomly time-variant linear channels. *IEEE Transactions on Communications Systems*, 11(4):360–393, December 1963.

[10] A. Bennatan, D. Burshtein, G. Caire, and S. Shamai. Superposition coding for side-information channels. *IEEE Transactions on Information Theory*, 52(5):1872–1889, May 2006.

[11] H. Bolcskei, M. Borgmann, and A.J. Paulraj. Space-frequency coded MIMO-OFDM with variable multiplexing-diversity tradeoff. In *Proceedings of IEEE International Conference on Communications (ICC'03)*, volume 4, pages 2837–2841, 11–15 May 2003.

[12] J.F. Borin and N.L.S. da Fonseca. Scheduler for IEEE 802.16 networks. *IEEE Communications Letters*, 12(4):274–276, April 2008.

[13] J. Boutros, G. Caire, E. Viterbo, H. Sawaya, and S. Vialle. Turbo code at 0.03 dB from capacity limit. In *Proceedings of 2002 IEEE International Symposium on Information Theory*, page 56, 2002.

[14] Isabella Cerutti, Filippo Meucci, Piero Castoldi, and Laura Pierucci. An adaptive cross-layer strategy for QoS-guaranteed links in 4G networks. In *IEEE Globecom 2008 Wireless Networking Symposium*, New Orleans, LA, USA, 2008.

[15] Isabella Cerutti, Luca Valcarenghi, Dania Marabissi, Filippo Meucci, Laura Pierucci, Luca Simone Ronga, Piero Castoldi, and Enrico Del Re. Enhanced operation modes in IEEE 802.16 and integration with optical MANs. In *Proceedings of the 12th Wireless Communications CNIT Thyrrenian Symposium*, 2007.

[16] Kyungwhoon Cheun. Performance of direct-sequence spread-spectrum RAKE receivers with random spreading sequences. *IEEE Transactions on Communications*, 45(9):1130–1143, September 1997.

[17] J. Choi, J. Yoo, and C.K. Kim. A distributed fair scheduling scheme with a new analysis model in IEEE 802.11 wireless LANs. *IEEE Transactions in Vehicular Technology*, 57(5):3083–3093, September 2008.

[18] C. Cicconetti, L. Lenzini, E. Mingozzi, and C. Eklund. Quality of service support in IEEE 802.16 networks. *IEEE Network*, 20(2):50–55, March–April 2006.

[19] M. Costa. Writing on dirty paper (corresp.). *IEEE Transactions on Information Theory*, 29(3):439–441, May 1983.

[20] J.W. Craig. A new, simple and exact result for calculating the probability of error for two-dimensional signal constellations. In *Proceedings of Military Communications Conference (MILCOM'91), Conference Record, 'Military Communications in a Changing World', IEEE*, volume 2, pages 571–575, November 1991.

[21] A.D. Dabbagh and D.J. Love. Multiple antenna MMSE based downlink precoding with quantized feedback or channel mismatch. *IEEE Transactions on Communications*, 56(11):1859–1868, November 2008.

[22] E. de Carvalho and P. Popovski. ARQ strategies for 2×2 spatially multiplexed MIMO systems. In *Proceedings of Fortieth Asilomar Conference on Signals, Systems and Computers (ACSSC'06)*, pages 1666–1670, 2006.

[23] E. de Carvalho and P. Popovski. Strategies for ARQ in 2X2 MIMO Systems. *IEEE Communications Letters*, 12(6):441–443, June 2008.

[24] Enrico Del Re. *Elements of Digital Signal Processing (Elementi di elaborazione numerica dei segnali)*. Pitagora Editrice, 1997.

[25] P. Dent, G.E. Bottomley, and T. Croft. Jakes fading model revisited. *Electronics Letters*, 29(13):1162–1163, June 1993.

[26] B. Muquet, E. Biglieri, A. Goldsmith, and H. Sari. MIMO link adaptation in mobile WiMAX systems. In *Proceedings WCNC 2007*, Hong Kong, March 2007.

[27] A. El Gamal and T.M. Cover. Multiple user information theory. *Proceedings of the IEEE*, 68(12):1466–1483, December 1980.

[28] S.-E. Elayoubi and B. Fourestie. Performance evaluation of admission control and adaptive modulation in OFDMA WiMax systems. *IEEE/ACM Transactions on Networking*, 16(5), October 2008.

[29] Emre Telatar. Capacity of multi-antenna Gaussian channels. *European Transactions on Telecommunications*, 10(6):585–595, 1999.

[30] Gerard J. Foschini. Layered space-time architecture for wireless communication in a fading environment when using multi-element antennas. *Bell Laboratories Technical Journal*, pages 41–59, 1996.

[31] G.J. Foschini and M.J. Gans. On limits of wireless communications in a fading environment when using multiple antennas. *Wireless Personal Communications*, 6(3):311–355, 1998.

[32] I-Kang Fu, Wendy C. Wong, David Chen, Peter Wang, Mike Hart, and Sunil Vadgama. Path-loss and shadow fading models for IEEE 802.16j Relay Task Group. Technical Report, IEEE 802.16 Relay Task Group, Document C80216j-06 045r1, July 2006.

[33] M.J. Gans, N. Amitay, Y.S. Yeh, Hao Xu, T.C. Damen, R.A. Valenzuela, T. Sizer, R. Storz, D. Taylor, W.M. MacDonald, Cuong Tran, and A. Adamiecki. Outdoor BLAST measurement system at 2.44 GHz: Calibration and initial results. *IEEE Journal on Selected Areas in Communications*, 20(3):570–583, April 2002.

[34] David Gesbert, Marios Kountouris, Robert W. Heath, Chan byoung Chae, and Thomas Slzer. From single user to multiuser communications: Shifting the MIMO paradigm. In *IEEE Signal Processing Magazine*, 24(5):36–46, October 2007.

[35] A. Goldsmith, S.A. Jafar, N. Jindal, and S. Vishwanath. Capacity limits of MIMO channels. *IEEE Journal on Selected Areas in Communications*, 21(5):684–702, June 2003.

[36] A.J. Goldsmith, L.J. Greenstein, and G.J. Foschini. Error statistics of real-time power measurements in cellular channels with multipath and shadowing. *IEEE Transactions on Vehicular Transactions*, 43(3):439–446, Part 1–2, August 1994.

[37] S. Gowrisankar and B.S. Rajan. A rate-one full-diversity low-complexity space-time-frequency block code (STFBC) for 4-Tx MIMO-OFDM. In *Proceedings of International Symposium on Information Theory (ISIT 2005)*, pages 2090–2094, September 2005.

[38] Donald Gross and Carl M. Harris. *Fundamentals of Queueing Theory*, 2nd ed. John Wiley & Sons, New York, NY, USA, 1985.

[39] M. Gudmundson. Correlation model for shadow fading in mobile radio systems. *Electronics Letters*, 27(23):2145–2146, November 1991.

[40] M. Guillaud, D.T.M. Slock, and R. Knopp. A practical method for wireless channel reciprocity exploitation through relative calibration. In *Proceedings Signal Processing and Its Applications*, volume 1, pages 403–406, 2005.

[41] R.W. Heath and A.J. Paulraj. Switching between diversity and multiplexing in MIMO systems. *IEEE Transactions on Communications*, 53(6):962–968, June 2005.

[42] Ari Hottinen, Olav Tirkkonen, and Risto Wichman. *Multi-Antenna Transceiver Techniques for 3G and Beyond*. John Wiley & Sons, 2003.

[43] D. Hwang, B. Clerckx, and G. Kim. Regularized channel inversion with quantized feedback in down-link multiuser channels. *IEEE Transactions on Wireless Communications*, 8(12):5785–5789, December 2009.

[44] IEEE 802.16 Broadband Wireless Access Working Group. IEEE 802.16m Evaluation Methodology Document, IEEE 802.16m-08/004r5, January 2009.

[45] IEEE 802.16 Broadband Wireless Access Working Group. IEEE 802.16m System Description Document (SDD) IEEE 802.16m-09/0034r2, September 2009.

[46] IEEE 802.16 Broadband Wireless Access Working Group. IEEE 802.16m System Requirements Document (SRD) IEEE 802.16m-07/002r9, September 2009.

[47] IEEE Std 802.16-2004 (Revision of IEEE Std 802.16-2001). IEEE Standard for Local and Metropolitan Area Networks Part 16: Air Interface for Fixed Broadband Wireless Access Systems, 2004.

[48] IEEE Std 802.16e-2005 and IEEE Std 802.16-2004 Cor 1-2005 (Amendment and Corrigendum to IEEE Std 802.16-2004). IEEE Standard for Local and metropolitan area networks Part 16: Air Interface for Fixed and Mobile Broadband Wireless Access Systems Amendment 2: Physical and Medium Access Control Layers for Combined Fixed and Mobile Operation in Licensed Bands and Corrigendum 1, 2006.

[49] ITU-R M.2134. Requirements related to technical performance for IMT-Advanced radio interface(s). Technical Report, ITU-R, 2008.

[50] H. Jafarkhani. A quasi-orthogonal space-time block code. *IEEE Transactions on Communications*, 49(1):1–4, January 2001.

[51] D.R.V. Jagannadha Rao, V. Shashidhar, Z.A. Khan, and B.S. Rajan. Low-complexity, full-diversity space-time-frequency block codes for MIMO-OFDM. In *IEEE Global Telecommunications Conference, GLOBECOM'04*, volume 1, pages 204–208, November–December 2004.

[52] Zhanjun Jiang, Wen Pan, Dongming Wang, and Xiaohu You. Study on capacity of BLAST system with error propagation. In *Proceedings of 4th IEEE International Conference on Circuits and Systems for Communications (ICCSC 2008)*, pages 387–391, 26–28 May 2008.

[53] Mingyu Kang, Young Jin Sang, Hae Gwang Hwang, Hyung Yeol Lee, and Kwang Soon Kim. Performance Analysis of Proportional Fair Scheduling with Partial Feedback Information for Multiuser Multicarrier Systems. In *Proceedings of IEEE 69th Vehicular Technology Conference VTC Spring 2009*, pages 1–5, 26–29 April 2009.

[54] Tae-Sung Kang and Hyung-Myung Kim. Optimal beam subset and user selection for orthogonal random beamforming. *IEEE Communications Letters*, 12(9):636–638, September 2008.

[55] Zhengjiu Kang, Kung Yao, and F. Lorenzelli. Nakagami-m fading modeling in the frequency domain for OFDM system analysis. *IEEE Communications Letters*, 7(10):484–486, October 2003.

[56] M.Z.A. Khan and B.S. Rajan. Space-time block codes from co-ordinate interleaved orthogonal designs. In *Proceedings of IEEE International Symposium on Information Theory*, page 275, 2002.

[57] Il-Min Kim, Seung-Chul Hong, S.S. Ghassemzadeh, and V. Tarokh. Opportunistic beamforming based on multiple weighting vectors. *IEEE Transactions on Wireless Communications*, 4(6):2683–2687, November 2005.

[58] Leonard Kleinrock. *Queueing Systems*. Wiley-Interscience, 1975.

[59] M. Kountouris and D. Gesbert. Memory-based opportunistic multi-user beamforming. In *Proceedings of International Symposium on Information Theory (ISIT2005)*, pages 1426–1430, September 4–9, 2005.

[60] Pekka Kysti. *Matlab SW documentation of WIM2 model*. WINNER, Wireless World Initiative New Radio, www.ist-winner.org, September 2008.

[61] Pekka Kysti et al. WINNER II channel models, D1.1.2. Technical Report, IST-4-027756 WINNER II, 2007.

[62] E. Laias, I. Awan, and P.M.L. Chan. Fair and latency aware uplink scheduler in ieee 802.16 using customized deficit round robin. In *International Conference on Advanced*

Information Networking and Applications Workshops (WAINA'09), pages 425–432, May 2009.

[63] E. Larsson and E. Jorswieck. Competition versus cooperation on the miso interference channel. *IEEE Journal on Selected Areas in Communications*, 26(7):1059–1069, September 2008.

[64] P. Larsson. Analysis of multi-user ARQ with multiple unicast flows under non-IID reception probabilities. In *Proceedings of IEEE Wireless Communications and Networking Conference*, pages 384–388, March 2007.

[65] P. Larsson and N. Johansson. Multi-User ARQ. In *IEEE 63rd Vehicular Technology Conference, VTC Spring 2006*, volume 4, pages 2052–2057, May 2006.

[66] A. Lera, A. Molinaro, and S. Pizzi. Channel-aware scheduling for QoS and fairness provisioning in IEEE 802.16/WiMAX broadband wireless access systems. *IEEE Network*, 21(5):34–41, 2007.

[67] Ye Li and L.J. Cimini, Jr. Bounds on the interchannel interference of OFDM in time-varying impairments. *IEEE Transactions on Communications*, 49(3):401–404, 2001.

[68] Xue-Bin Liang and Xiang-Gen Xia. On the nonexistence of rate-one generalized complex orthogonal designs. *IEEE Transactions on Information Theory*, 49(11):2984–2988, November 2003.

[69] Qingwen Liu, Xin Wang, and G.B. Giannakis. A cross-layer scheduling algorithm with QoS support in wireless networks. *IEEE Transactions on Vehicular Technology*, 55(3):839–847, May 2006.

[70] Qingwen Liu, Shengli Zhou, and G.B. Giannakis. Queuing with adaptive modulation and coding over wireless links: cross-layer analysis and design. *IEEE Transactions on Wireless Communications*, 4(3):1142–1153, May 2005.

[71] Zhiqiang Liu, Yan Xin, and G.B. Giannakis. Space-time-frequency coded OFDM over frequency-selective fading channels. *IEEE Transactions on Signal Processing*, 50(10):2465–2476, October 2002.

[72] Songwu Lu, V. Bharghavan, and R. Srikant. Fair scheduling in wireless packet networks. *IEEE/ACM Transactions on Networking*, 7(4):473–489, August 1999.

[73] A. Maaref and S. Aissa. Performance analysis of orthogonal space-time block codes in spatially correlated MIMO Nakagami fading channels. *IEEE Transactions on Wireless Communications*, 5(4):807–817, April 2006.

[74] Dania Marabissi, Filippo Meucci, Laura Pierucci, and Luca Simone Ronga. Adaptive selection of MIMO schemes in IEEE 802.16e. In *IWCMC'07: Proceedings of the 2007 International Conference on Wireless Communications and Mobile Somputing*, pages 423–428, ACM, New York, NY, USA, 2007.

[75] D. Marco and M. Effros. On lossless coding with coded side information. *IEEE Transactions on Information Theory*, 55(7):3284–3296, July 2009.

[76] M.R. McKay, I.B. Collings, A. Forenza, and R.W. Heath. Multiplexing/beamforming switching for coded MIMO in spatially correlated channels based on closed-form BER approximations. *IEEE Transactions on Vehicular Technology*, 56(5):2555–2567, September 2007.

[77] R. Mersereau and T. Speake. The processing of periodically sampled multidimensional signals. *IEEE Transactions on Acoustics, Speech, and Signal Processing*, 31(1):188–194, February 1983.

[78] Fiilippo Meucci, Elisabeth De Carvalho, and Petar Popovski. Multi-user MIMO ARQ strategies. *IEEE Trans. Wireless Commun.*, submitted, 2009.

[79] Filippo Meucci, Orlando Cabral, Albena Mihovska, Fernando Velez, and Neeli Prasad. Spectrum aggregation with multi-band user allocation over two frequency bands. In *IEEE Mobile WiMAX Symposium*, Napa Valley, California, USA, 2009.

[80] Filippo Meucci, Albena Mihovska, Bayu Anggorojati, and Neeli Prasad. Joint resource allocation and admission control mechanism for an OFDMA-Based system. In *The 11th International Symposium on Wireless Personal Multimedia Communications*, Saariselka, Lapland, Finland, 2008.

[81] Filippo Meucci, Laura Pierucci, Enrico Del Re, and Ramjee Prasad. Dynamic allocation of STBC for OFDM based on WSSUS channel model. In *Wireless Communication Society, Vehicular Technology, Information Theory and Aerospace & Electronics Systems Technology*, Aalborg, Denmark, 2009.

[82] Filippo Meucci, Satya A. Wardana, and Neeli Rashmi Prasad. Secure physical layer using dynamic permutations in cognitive OFDMA systems. In *Proceedings of the IEEE 69th Vehicular Technology Conference, VTC2009-Spring*, Barcellona, 2009.

[83] A.F. Molisch, M.Z. Win, and J.H. Winters. Space-time-frequency (STF) coding for MIMO-OFDM systems. *IEEE Communications Letters*, 6(9):370–372, September 2002.

[84] M. Moustafa. Switched-mode BLAST technique for MIMO communications. In *Proceedings 11th International Conference on Advanced Communication Technology ICACT 2009*, volume 3, pages 1499–1502, 15–18 February 2009.

[85] Loutfi Nuaymi. *WiMAX: Technology for Broadband Wireless Access*. Wiley Inter-Science, January 2007.

[86] P. Piggin. Emerging mobile WiMAX antenna technologies. *IET Communications Engineer*, 4(5):29–33, 2006.

[87] Li Ping and Peng Wang. Multi-user gain and maximum eigenmode beamforming for mimo systems with rate constraints. In *IEEE Information Theory Workshop on Information Theory for Wireless Networks*, pages 1–5, July 2007.

[88] Petar Poposki. An alternative viewpoint on the (cross)layering in wireless networks. Unpublished Draft, 2006.

[89] Ramjee Prasad. A perspective of layerless communications. *Wireless Personal Communications*, 44(1):95–100, 2008.

[90] John G. Proakis. *Digital Communications*, 4th ed. McGraw-Hill, 1995.

[91] J. Radon. *Lineare scharen orthogonaler Matrizen*. Abhandlungen aus dem Mathematischen Seminar der Hamburgischen Universitat, 1922.

[92] K. Ramasubramanian and K. Baum. An OFDM timing recovery scheme with inherent delay-spread estimation. In *Proceedings IEEE Global Telecommunications Conference GLOBECOM'01*, volume 5, pages 3111–3115, 25–29 November 2001.

[93] C. Rao and B. Hassibi. Diversity-multiplexing gain trade-off of a MIMO system with relays. In *IEEE Information Theory Workshop on Information Theory for Wireless Networks*, pages 1–5, July 2007.

[94] J. Razavilar, K.J.R. Liu, and S.I. Marcus. Jointly optimized bit-rate/delay control policy for wireless packet networks with fading channels. *IEEE Transactions on Communications*, 50(3):484–494, March 2002.

[95] Alexander Sayenko, Vitaliy Tykhomyrov, Henrik Martikainen, and Olli Alanen. Performance analysis of the IEEE 802.16 ARQ mechanism. In *MSWiM'07: Proceedings of the 10th ACM Symposium on Modeling, analysis, and simulation of wireless and mobile systems*, pages 314–322, ACM, New York, NY, USA, 2007.

[96] C.E. Shannon. A mathematical theory of communication. *The Bell System Technical Journal*, 27:379–423 and 623–656, 1948.

[97] Hyundong Shin and Jae Hong Lee. Performance analysis of space-time block codes over keyhole Nakagami-m fading channels. *IEEE Transactions on Vehicular Transactions*, 53(2):351–362, March 2004.

[98] Marvin K. Simon and Mohamed-Slim Alouini. *Digital Communication over Fading Channels*, 2nd ed. Wiley/IEEE Press, 2005.

[99] Chakchai So-In, R. Jain, and A.-K. Tamimi. Scheduling in IEEE 802.16e mobile WiMAX networks: Key issues and a survey. *IEEE Journal on Selected Areas in Communications*, 27(2):156–171, 2009.

[100] T.B. Sorensen, P.E. Mogensen, and F. Frederiksen. Extension of the ITU channel models for wideband (OFDM) systems. In *Vehicular Technology Conference, 2005. VTC-2005-Fall. 2005 IEEE 62nd*, volume 1, pages 392–396, September 2005.

[101] A. Soysal and S. Ulukus. Optimality of beamforming in fading MIMO multiple access channels. *IEEE Transactions on Communications*, 57(4):1171–1183, April 2009.

[102] G.L. Stuber, J.R. Barry, S.W. McLaughlin, Ye Li, M.A. Ingram, and T.G. Pratt. Broadband MIMO-OFDM wireless communications. *Proceedings of the IEEE*, 92(2):271–294, February 2004.

[103] Li Tan, Gang Su, Guangxi Zhu, and Peng Shang. Adaptive proportional fair scheduling based on opportunistic beamforming for MIMO systems. In *Proceedings of 5th International Conference on Wireless Communications, Networking and Mobile Computing WiCom'09*, pages 1–4, 24–26 September 2009.

[104] V. Tarokh, H. Jafarkhani, and A.R. Calderbank. Correction to space-time block codes from orthogonal designs. *IEEE Transactions on Information Theory*, 46(1):314–314, January 2000.

[105] V. Tarokh, H. Jafarkhani, and A.R. Calderbank. Space-time block codes from orthogonal designs. *IEEE Transactions on Information Theory*, 45(5):1456–1467, July 1999.

[106] V. Tarokh, N. Seshadri, and A.R. Calderbank. Space-time codes for high data rate wireless communication: performance criterion and code construction. *IEEE Transactions on Information Theory*, 44(2):744–765, March 1998.

[107] L. Tassiulas and S. Sarkar. Maxmin fair scheduling in wireless networks. In *Proceedings of IEEE Twenty-First Annual Joint Conference of the IEEE Computer and Communications Societies INFOCOM 2002*, volume 2, pages 763–772, 23–27 June 2002.

[108] L. Tassiulas and S. Sarkar. Maxmin fair scheduling in wireless ad hoc networks. *IEEE Journal on Selected Areas in Communications*, 23(1):163–173, January 2005.

[109] Ragnar Thobaben. Joint network/channel coding for multi-user hybrid-ARQ. In *7th International ITG Conference on Source and Channel Coding*, 2008.

[110] S.C. Thompson, A.U. Ahmed, J.G. Proakis, J.R. Zeidler, and M.J. Geile. Constant envelope OFDM. *IEEE Transactions on Communications*, 56(8):1300–1312, August 2008.

[111] Chao Tian and S.N. Diggavi. Side-information scalable source coding. *IEEE Transactions on Information Theory*, 54(12):5591–5608, December 2008.

[112] M. Tran, A. Doufexi, and A. Nix. Mobile WiMAX MIMO performance analysis: Downlink and uplink. In *Proceedings of PIMRC*, pages 1–5, September 2008.

[113] D.N.C. Tse, P. Viswanath, and Lizhong Zheng. Diversity-multiplexing tradeoff in multiple-access channels. *IEEE Transactions on Information Theory*, 50(9):1859–1874, September 2004.

[114] Richard van Nee and Ramjee Prasad. *OFDM for Wireless Multimedia Communications*. Artech House, Norwood, MA, USA, 2000.

[115] A. Vielmon, Ye Li, and J.R. Barry. Performance of Alamouti transmit diversity over time-varying Rayleigh-fading channels. *IEEE Transactions on Wireless Communications*, 3(5):1369–1373, September 2004.

[116] P. Viswanath, D.N.C. Tse, and R. Laroia. Opportunistic beamforming using dumb antennas. *IEEE Transactions on Information Theory*, 48(6):1277–1294, June 2002.

[117] Xin Wang, Qingwen Liu, and G.B. Giannakis. Analyzing and optimizing adaptive modulation-coding jointly with ARQ for QoS-guaranteed traffic. In *IEEE International Conference on Communications (ICC'06)*, volume 3, pages 1008–1013, June 2006.

[118] K.K. Wee and S.W. Lee. Priority based bandwidth allocation scheme for WIMAX systems. In *Proceedings of 2nd IEEE International Conference on Broadband Network & Multimedia Technology (IC-BNMT'09)*, pages 15–18, 18–20 October 2009.

[119] Jyh-Horng Wen, Shu-Hong Lee, Gwo-Ruey Lee, and Jin-Tong Chang. Timing and delay spread estimation scheme in OFDM systems. *IEEE Transactions on Consumer Electronics*, 54(2):316–320, May 2008.

[120] M.A. Wigger and G. Kramer. Three-user MIMO MACs with cooperation. In *IEEE Information Theory Workshop on Networking and Information Theory (ITW 2009)*, pages 221–225, June 2009.

[121] K. Witrisal, Yong-Ho Kim, and R. Prasad. A new method to measure parameters of frequency-selective radio channels using power measurements. *IEEE Communications Letters*, 49(10):1788–1800, October 2001.

[122] P.W. Wolniansky, G.J. Foschini, G.D. Golden, and R.A. Valenzuela. V-BLAST: An architecture for realizing very high data rates over the rich-scattering wireless channel. In *Proceedings of 1998 URSI International Symposium on Signals, Systems, and Electronics (ISSSE 98)*, pages 295–300, 29 September–2 October 1998.

[123] Kai-Kit Wong, R.D. Murch, and K.B. Letaief. Performance enhancement of multiuser MIMO wireless communication systems. *IEEE Transactions on Communications*, 50(12):1960–1970, December 2002.

[124] Pengfei Xia, Shengli Zhou, and G.B. Giannakis. Adaptive MIMO-OFDM based on partial channel state information. *IEEE Transactions on Signal Processing* [see also *IEEE Transactions on Acoustics, Speech, and Signal Processing*], 52(1):202–213, January 2004.

[125] Chengshan Xiao and Y.R. Zheng. Transmit precoding for MIMO systems with partial CSI and discrete-constellation inputs. In *Proceedings of IEEE International Conference on Communications ICC'09*, pages 1–5, June 14–18, 2009.

[126] Ning Xu, Shaohua Yu, and Xueshun Wang. A combined fair scheduling algorithm for combined-input-crosspoint-queued switch. In *Proceedings of 5th International Confer-*

ence on Wireless Communications, Networking and Mobile Computing (WiCom'09), pages 1–4, 24–26 September 2009.

[127] Hassan Yaghoobi. Scalable OFDMA physical layer in IEEE 802.16 WirelessMAN. *Intel Technology Journal*, 8(3):201–212, 2004.

[128] Zhiwei Yan, Lei Huang, and C.-C.J. Kuo. Seamless high-velocity handover support in mobile WiMAX networks. In *Proceedings of IEEE Singapore International Conference on Communication Systems*, pages 1680–1684, November 2008.

[129] T. Yucek and H. Arslan. Time dispersion and delay spread estimation for adaptive OFDM systems. *IEEE Transactions on Vehicular Transactions*, 57(3):1715–1722, May 2008.

[130] Li Zhang, A. Burr, M. Darnell, and Lingyang Song. Performance evaluation of spatial multiplexing systems over finite scattering channel. In *Proceedings of ITW'06 Chengdu Information Theory Workshop IEEE*, pages 658–662, 22–26 October 2006.

[131] W. Zhang, X.-G. Xia, and P.C. Ching. High-rate full-diversity space-time-frequency codes for broadband MIMO block-fading channels. *IEEE Transactions on Communications*, 55(1):25–34, January 2007.

[132] Haitao Zheng, Angel Lozano, and Mohamed Haleem. Multiple ARQ processes for MIMO systems. *EURASIP Journal on Applied Signal Processing*, 2004(5):772–782, 2004.

[133] L. Zheng and D.N.C. Tse. Sphere packing in the Grassmann manifold: A geometric approach to the noncoherent multi-antenna channel. In *Proceedings of IEEE International Symposium on Information Theory*, page 364, 25–30 June 2000.

[134] Lizhong Zheng and D.N.C. Tse. Diversity and multiplexing: A fundamental tradeoff in multiple-antenna channels. *IEEE Transactions on Information Theory*, 49(5):1073–1096, May 2003.

[135] Xiangyang Zhuang, F.W. Vook, S. Rouquette-Leveil, and K. Gosse. Transmit diversity and spatial multiplexing in four-transmit-antenna OFDM. In *Proceedings of IEEE ICC*, volume 4, pages 2316–2320, 11–15 May 2003.

Index

About the Author

Filippo Meucci received the M.S. degree in Electronics Engineering cum laude et mentione in 2005 from the University of Florence, Italy jointly with University of California, Berkeley. In December 2009 he received his Ph.D. in Telecommunication Engineering from the University of Florence, Italy.

From 2005–2007 he has been working with Italian Highways Company for high-speed vehicle radio-localization and on-vehicle broadband delivery. He also worked as a Ph.D. fellow at Aalborg University from January 2008 to May 2009. Currently, he is a research fellow at University of Florence and Selex Galileo.

He is co-founder of Engineering without Borders, Florence, where he has held the position of steering committee chair for several years and he has managed several ICT-focused international cooperation projects.

His research interests include: broadband MIMO-OFDM wireless networks, adaptive PHY and MAC layers, cross-layer designs, MIMO radars and appropriate ICT technologies for human development.

RIVER PUBLISHERS SERIES IN COMMUNICATIONS

Volume 1
4G Mobile & Wireless Communications Technologies
Sofoklis Kyriazakos, Ioannis Soldatos, George Karetsos
September 2008
ISBN: 978-87-92329-02-8

Volume 2
Advances in Broadband Communication and Networks
Johnson I. Agbinya, Oya Sevimli, Sara All, Selvakennedy Selvadurai, Adel Al-Jumaily, Yonghui Li, Sam Reisenfeld
October 2008
ISBN: 978-87-92329-00-4

Volume 3
Aerospace Technologies and Applications for Dual Use A New World of Defense and Commercial in 21st Century Security
General Pietro Finocchio, Ramjee Prasad, Marina Ruggieri
November 2008
ISBN: 978-87-92329-04-2

Volume 4
Ultra Wideband Demystified Technologies, Applications, and System Design Considerations
Sunil Jogi, Manoj Choudhary
January 2009
ISBN: 978-87-92329-14-1

Volume 5
Single- and Multi-Carrier MIMO Transmission for Broadband Wireless Systems
Ramjee Prasad, Muhammad Imadur Rahman, Suvra Sekhar Das, Nicola Marchetti
April 2009
ISBN: 978-87-92329-06-6

Volume 6
Principles of Communications: A First Course in Communications
Kwang-Cheng Chen
June 2009
ISBN: 978-87-92329-10-3

Volume 7
Link Adaptation for Relay-Based Cellular Networks
Başak Can
November 2009
ISBN: 978-87-92329-30-1

Volume 8
Planning and Optimisation of 3G and 4G Wireless Networks
J.I. Agbinya
January 2010
ISBN: 978-87-92329-24-0

Volume 9
Towards Green ICT
Ramjee Prasad, Shingo Ohmori, Dina Šimunić
June 2010
ISBN: 978-87-92329-38-7

Volume 10
Adaptive PHY-MAC Design for Broadband Wireless Systems
Ramjee Prasad, Suvra Sekhar Das, Muhammad Imadur Rahman
August 2010
ISBN: 978-87-92329-08-0

Volume 11
Multihop Mobile Wireless Networks
Kannan Govindan, Deepthi Chander, Bhushan G. Jagyasi, Shabbir N. Merchant, Uday B. Desai
October 2010
ISBN: 978-87-92329-44-8

Volume 12
Telecommucations in Disaster Areas
Nicola Marchetti
November 2010
ISBN: 978-87-92329-48-6

www.ingramcontent.com/pod-product-compliance
Ingram Content Group UK Ltd.
Pitfield, Milton Keynes, MK11 3LW, UK
UKHW021120180425
457613UK00005B/159

* 9 7 8 8 7 9 2 3 2 9 5 0 9 *